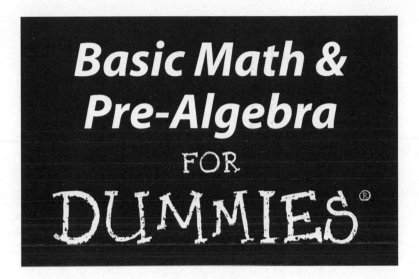

Basic Math & Pre-Algebra

FOR DUMMIES®

by Mark Zegarelli

BICENTENNIAL
1807
WILEY
2007
BICENTENNIAL

Wiley Publishing, Inc.

Basic Math & Pre-Algebra For Dummies®

Published by
Wiley Publishing, Inc.
111 River St.
Hoboken, NJ 07030-5774
www.wiley.com

For general information on our other products and services, please contact our Customer Care Department within the U.S. at 877-762-2974, outside the U.S. at 317-572-3993, or fax 317-572-4002.

For technical support, please visit www.wiley.com/techsupport.

Wiley also publishes its books in a variety of electronic formats. Some content that appears in print may not be available in electronic books.

Library of Congress Control Number: 2007931551

ISBN: 978-0-470-13537-2

Manufactured in the United States of America

10 9 8

WILEY

About the Author

Mark Zegarelli is the author of *Logic For Dummies* (Wiley). He holds degrees in both English and math from Rutgers University. He has earned his living for many years writing vast quantities of logic puzzles, a hefty chunk of software documentation, and the occasional book or film review. Along the way, he's also paid a few bills doing housecleaning, decorative painting, and (for ten hours) retail sales. He likes writing best, though.

Mark lives mostly in Long Branch, New Jersey, and sporadically in San Francisco, California.

Dedication

I dedicate this book to the memory of my mother, Sally Ann Zegarelli (Joan Bernice Hanley).

Author's Acknowledgments

This is my second *For Dummies* book, and it's still a refreshing and inspiring experience to be surrounded by a team of talented folks committed to making both this book and my experience as a writer truly extraordinary. Many thanks for the editorial guidance and wisdom of Lindsay Lefevere, Natalie Harris, Danielle Voirol, and Sarah Faulkner of Wiley Publications.

Many thanks to Mrs. Beardsley of Wanamassa School and Mr. Mundy of Long Branch High School for starting me off right in math by making a good education straightforward and often fun. Thanks to Martin Gardner and his "Mathematical Games" column in *Scientific American* for showing me what amazing things numbers really are. Thanks to my math professors at Brookdale Community College: Barbara Tozzi, Eugene Jermael "Teach" Bowen, and Greg Liano. And thanks to all my professors at Rutgers University, especially Holly Carley, Zheng-Chao Han, Richard Lyons, and David Nacin for their support and encouragement.

And as always, thanks to my partner, Mark Dembrowski, for unfailing love and encouragement; my sister, Tami Pantella, for believing in me even when I don't; my cuz, Ed Bremer, for being a good pal and fellow math dude; and to my old, old, old, old, old friends Chip DeCraene, David Feaster, Michael Konopko, Rick Kawala, Brian London, Mark O'Malley, Stanley Marcus, Tim O'Rourke, Robert Rubin, Alison Sigethy, and Ken Wolfe.

And thanks to Maxfield's House of Caffeine in San Francisco for brewing the coffee.

Publisher's Acknowledgments

We're proud of this book; please send us your comments through our Dummies online registration form located at www.dummies.com/register/.

Some of the people who helped bring this book to market include the following:

Acquisitions, Editorial, and Media Development

Project Editor: Natalie Faye Harris

Acquisitions Editor: Lindsay Lefevere

Copy Editor: Danielle Voirol

Technical Editor: Carol Spilker

Editorial Manager: Christine Meloy Beck

Editorial Assistants: Leeann Harney, David Lutton, Erin Calligan Mooney, Joe Niesen

Cartoons: Rich Tennant (www.the5thwave.com)

Composition Services

Project Coordinator: Heather Kolter, Kristie Rees

Layout and Graphics: Carrie A. Cesavice, Denny Hager, Joyce Haughey, Shane Johnson, Stephanie D. Jumper, Christine Williams

Anniversary Logo Design: Richard Pacifico

Proofreaders: Laura L. Bowman, John Greenough, Susan Moritz, Christy Pingleton

Indexer: Slivoskey Indexing Services

Special Help
Sarah Faulkner

Publishing and Editorial for Consumer Dummies

> **Diane Graves Steele,** Vice President and Publisher, Consumer Dummies

> **Joyce Pepple,** Acquisitions Director, Consumer Dummies

> **Kristin A. Cocks,** Product Development Director, Consumer Dummies

> **Michael Spring,** Vice President and Publisher, Travel

> **Kelly Regan,** Editorial Director, Travel

Publishing for Technology Dummies

> **Andy Cummings,** Vice President and Publisher, Dummies Technology/General User

Composition Services

> **Gerry Fahey,** Vice President of Production Services

> **Debbie Stailey,** Director of Composition Services

Contents at a Glance

Table of Contents

Introduction

. . . .

*O*nce upon a time, you loved numbers. This isn't the first line of a fairy tale. Once upon a time, you really did love numbers. Remember?

Maybe you were 3 and your grandparents were visiting. You sat next to them on the couch and recited the numbers from 1 to 10. They were proud of you and — be honest — you were proud of yourself, too. Or maybe you were 5 and discovering how to write numbers, trying hard not to get your 6s and 7s backwards.

Learning was fun. *Numbers* were fun. So what happened? Maybe the trouble started with long division. Or sorting out how to change fractions to decimals. Could it have been figuring out how to add 8 percent sales tax to the cost of a purchase? Reading a graph? Converting miles to kilometers? Trying to find that most dreaded value of *x?* Wherever it started, you began to suspect that math didn't like you — and you didn't like math very much, either.

Why do people often enter preschool excited about learning how to count and leave high school as young adults convinced that they can't do math? The answer to this question would probably take 20 books this size, but solving the problem can begin right here.

I humbly ask you to put aside any doubts. Remember, just for a moment, an innocent time — a time before math inspired panic attacks or, at best, induced irresistible drowsiness. In this book, I take you from an understanding of the basics to the place where you're ready to enter any algebra class and succeed.

About This Book

Somewhere along the road from counting to algebra, most people experience the Great Math Breakdown. This feels something like when your car begins smoking and sputtering on a 110-degree highway somewhere between Noplace and Not Much Else!

Please consider this book your personal roadside helper, and think of me as your friendly math mechanic (only much cheaper!). Stranded on the inter-state, you may feel frustrated by circumstances and betrayed by your vehi-cle, but for the guy holding the toolbox, it's all in a day's work. The tools for fixing the problem are in this book.

Not only does this book help you with the basics of math, but it also helps you get past any aversion you may feel toward math in general! I've broken down the concepts into easy-to-understand sections. And because *Basic Math & Pre-Algebra For Dummies* is a reference book, you don't have to read the chapters or sections in order — you can look over only what you need. So feel free to jump around. Whenever I cover a topic that requires information from earlier in the book, I refer you to that section or chapter in case you want to refresh yourself on the basics.

Here are two pieces of advice I give all the time — remember them as you work your way through the concepts in this book:

✔ **Take frequent study breaks.** Every 20 to 30 minutes, stand up and push in your chair. Then feed the cat, do the dishes, take a walk, juggle tennis balls, try on last year's Halloween costume — do *something* to distract yourself for a few minutes. You'll come back to your books more productive than if you were to just sit there hour after hour with your eyes glazing over.

✔ **After you've read through an example and think you understand it, copy the problem, close the book, and try to work it through.** If you get stuck, steal a quick look — but later, try that same example again to see whether you can get through without opening the book. (Remember that on any tests you're preparing for, peeking is probably not allowed!)

Conventions Used in This Book

To help you navigate your way through this book, I use the following conventions:

✔ *Italicized* text highlights new words and defined terms.

✔ **Boldfaced** text indicates keywords in bulleted lists and the action part of numbered steps.

✔ `Monofont` text highlights Web addresses.

✔ Variables, such as x and y, are in italics.

What You're Not to Read

Although every author secretly (or not-so-secretly) believes that each word he pens is pure gold, you don't have to read every word in this book unless you really want to. Feel free to skip over sidebars (those shaded gray boxes) where I go off on a tangent — unless you find tangents interesting, of course. Paragraphs labeled with the Technical Stuff icon are also nonessential.

Foolish Assumptions

If you're planning to read this book, you're likely

- ✔ A student who wants a solid understanding of the basics of math for a class or test you're taking
- ✔ An adult who wants to improve skills in arithmetic, fractions, decimals, percentages, weights and measures, geometry, algebra, and so on for when you have to use math in the real world
- ✔ Someone who wants a refresher so you can help another person understand math

My only assumption about your skill level is that you can add, subtract, multiply, and divide. So to find out whether you're ready for this book, take this simple test:

$5 + 6 = $ ___

$10 - 7 = $ ___

$3 \times 5 = $ ___

$20 \div 4 = $ ___

If you can answer these four questions, you're ready to begin.

How This Book Is Organized

This book is organized into six parts, starting you off at the very beginning of math — with topics such as counting and the number line — and taking you all the way into algebra.

Part 1: Arming Yourself with the Basics of Basic Math

In Part I, I take what you already know about math and put it in perspective.

Chapter 1 gives you a brief history of what numbers are and where they came from. I discuss how number sequences arise. I show you how important sets of numbers — such as the counting numbers, the integers, and the rational numbers — all fit together on the number line. I also show you how to use the number line to perform basic arithmetic.

Chapter 2 discusses how digits are the building blocks of numbers, much the same as letters are the building blocks of words. I show you how the number system you use every day — the Hindu-Arabic number system (also called *decimal numbers*) — uses place value based on the number 10 to build digits into numbers.

Finally, Chapter 3 focuses on what I call the Big Four operations — adding, subtracting, multiplying, and dividing. I give you a refresher on how to do column addition with carrying, subtraction with borrowing, multiplication of large numbers, and the ever-dreaded long division.

Part II: Getting a Handle on Whole Numbers

In Part II, you take a leap forward in your understanding of how to work with the Big Four operations. In Chapter 4, I cover inverse operations, the commutative, associative, and distributive properties, and working with negative numbers. You discover how to use inequalities such as greater than (>) and less than (<). I also introduce you to more-advanced operations, such as powers (exponents), square roots, and absolute value.

Chapter 5 introduces what I call the *three Es of math:* expressions, equations, and evaluation. The rest of the chapter focuses on an all-important skill: evaluating mathematical expressions using the order of operations. In Chapter 6, you discover how to solve *word problems* (also called *story problems*) by creating word equations.

Chapter 7 takes a detailed look at divisibility. I show you a bunch of tricks for finding out whether one number is divisible by another. I also discuss prime numbers and composite numbers. Finally, Chapter 8 discusses factors and multiples, showing you how these two concepts are connected. I show you how to decompose a number to its prime factors. I also show you how to find the greatest common factor (GCF) and least common multiple (LCM) of two or more numbers.

Part III: Parts of the Whole: Fractions, Decimals, and Percents

Part III focuses on how mathematics represents parts of the whole as fractions, decimals, and percents and on how these three ideas are connected.

Chapters 9 and 10 focus on fractions, including increasing and reducing the terms of fractions. From there, I show you how to multiply and divide fractions, plus a variety of ways to add and subtract fractions. Finally, I show you how to work with mixed numbers. In Chapter 11, the topic is decimals. I show you how to add, subtract, multiply, and divide decimals and how to convert fractions to decimals and vice versa. I also give you an understanding of repeating decimals.

Chapter 12 discusses percentages. I show you how to convert percentages to both fractions and decimals, and vice versa. I follow up with a variety of ways for finding percentages, including a simple but powerful tool called the percent circle. Finally, Chapter 13 focuses on solving word problems that involve fractions, decimals, and percents.

Part IV: Picturing and Measuring — Graphs, Measures, Stats, and Sets

Part IV contains a variety of topics that build upon the skills you gain in the first three parts of the book.

In Chapter 14, I show you how scientific notation makes very large and very small numbers more manageable by combining decimals and powers of ten. In Chapter 15, I discuss two important systems of weights and measures: the English system (used mostly in the United States) and the metric system (used throughout the world). I give you a variety of conversion equations and show you how to convert units of measurement. I also give you a few rules of thumb for estimating metric units.

Chapter 16 discusses geometry, giving you a variety of formulas to find the perimeter and area of basic shapes and the surface area and volume of a few important solids.

Chapter 17 introduces you to graphing, first showing you three important types of graphs — the bar graph, the pie chart, and the line graph. I also give you the basics of the most common graphing method used in math, the Cartesian graph. I show you how to plot points, draw lines, and solve problems using this graph. Chapter 18 gives you still more practice solving word problems, especially those focusing on geometry and on weights and measures.

Chapter 19 introduces you to statistics and probability. You discover the difference between qualitative data and quantitative data and how to calculate both the mean average and the median average. I also show you how to calculate probability by counting possible outcomes and favorable outcomes.

In Chapter 20, I give you the basics of set theory, including how to define a set, identify elements and subsets, and understand the empty set. I also show you how to perform some basic operations on sets, including the union and the intersection of sets.

Part V: The X-Files: Introduction to Algebra

Part V is your introduction to algebra. Chapter 21 provides an overview of algebra, showing you the basics of declaring a variable (such as *x*) and then using it in expressions, making connections with what you already know about expressions from Chapter 5.

Chapter 22 gives you a variety of ways to solve algebraic equations. Finally, Chapter 23 puts it all together, showing you how to solve algebra word problems from beginning to end.

Part VI: The Part of Tens

Just for fun, this part of the book includes a few top-ten lists on a variety of topics, including key math concepts and number sets.

Icons Used in This Book

Throughout the book, I use four icons to highlight what's hot and what's not:

This icon points out key ideas that you need to know. Make sure you understand before reading on! Remember this info even after you close the book.

Tips are helpful hints that show you the quick and easy way to get things done. Try them out, especially if you're taking a math course.

Warnings flag common errors that you want to avoid. Get clear about where these little traps are hiding so you don't fall in.

This icon points out interesting trivia that you can read or skip over as you like.

Where to Go from Here

You can use this book in a few ways. If you're reading this book without imme-diate time pressure from a test or homework assignment, you can certainly start at the beginning and keep on going through to the end. The advantage to this method is that you realize how much math you *do* know — the first few chapters go very quickly. You gain a lot of confidence as well as some practi-cal knowledge that can help you later on, because the early chapters also set you up to understand what follows.

Or how about this: When you're ready to work, read up on the topic you're studying. Leave the book on your nightstand and, just before bed, spend a few minutes reading the easy stuff from the early chapters. You'd be surprised how a little refresher on simple stuff can suddenly cause more-advanced concepts to click.

If your time is limited — especially if you're taking a math course and you're looking for help with your homework or an upcoming test — skip directly to the topic you're studying. Wherever you open the book, you can find a clear explanation of the topic at hand, as well as a variety of hints and tricks. Read through the examples and try to do them yourself, or use them as templates to help you with assigned problems.

Here's a short list of topics that tend to back students up:

- ✔ Negative numbers (Chapter 4)
- ✔ Order of operations (Chapter 5)
- ✔ Word problems (Chapters 6, 13, 18, and 23)
- ✔ Factoring numbers (Chapter 8)
- ✔ Fractions (Chapters 9 and 10)

Most of these topics are in Parts II and III, but they're foundational to what's cov-ered later in the book. Generally speaking, any time you spend building these five skills is like money in the bank as you proceed in math, so you may want to visit these sections several times. As soon as you feel comfortable adding nega-tive numbers or multiplying fractions, your confidence soars. Moreover, every-thing that I cover in the rest of the book should seem much simpler.

If you get stuck along the way, take a break and come back to the problem; you may find that the answer suddenly clicks for you as soon as you read it again with a refreshed mind. If you still feel stuck, you may need to drop back a few pages and read from the beginning of the section or chapter. Sometimes, work-ing through a few easier examples is the best way to prepare for when the going gets tough.

Part I

Arming Yourself with the Basics of Basic Math

The 5th Wave   By Rich Tennant

I'm mathematically dyslexic. But it's not that unusual— 100 out of every 15 people are.

In this part . . .

You already know more about math than you think you know. Here, you review and gain perspective on basic math ideas such as number patterns, the number line, how place value based on the number ten turns digits into numbers, and how zero functions as a place-holder. I also reintroduce you to what I call the Big Four operations (adding, subtracting, multiplying, and dividing).

Chapter 1

Playing the Numbers Game

One of the useful things about numbers is that they're *conceptual,* which means that in an important sense, they're all in your head. (This fact, however, probably won't get you out of having to know about them — nice try!)

For example, you can picture three of anything: three cats, three baseballs, three cannibals, three planets. But just try to picture the concept of three all by itself, and you find it's impossible. Oh, sure, you can picture the numeral 3, but the *threeness* itself — much like love or beauty or honor — is beyond direct understanding. But after you have the *concept* of three (or four, or a million), you have access to an incredibly powerful system for understanding the world: mathematics.

In this chapter, I give you a brief history of how numbers came into being. I discuss a few common *number sequences* and show you how these connect with simple math *operations* like addition, subtraction, multiplication, and division.

After that, I describe how some of these ideas come together with a simple yet powerful tool – the *number line*. I discuss how numbers are arranged on the number line, and I also show you how to use the number line as a calculator for simple arithmetic.

Finally, I describe how the *counting numbers* (1, 2, 3, . . .) sparked the invention of more unusual types of numbers such as *negative numbers, fractions*, and *irrational numbers*. I also show you how these *sets of numbers* are *nested* — that is, how one set of numbers fits inside another, which fits inside another.

Inventing Numbers

Historians believe that the first number systems came into being at the same time as agriculture and commerce. Before that, people in prehistoric, hunter-gatherer societies were pretty much content to identify bunches of things as "a lot" or "a little."

But as farming developed and trade between communities began, more precision was needed. So people began using stones, clay tokens, and similar objects to keep track of their goats, sheep, oil, grain, or whatever commodity they had. These tokens could be exchanged for the objects they represented in a one-to-one exchange.

Eventually, traders realized that they could draw pictures instead of using tokens. Those pictures evolved into tally marks and, in time, into more complex systems. Whether they realized it or not, their attempts to keep track of commodities had led these early humans to invent something entirely new: *numbers*.

Throughout the ages, the Babylonians, Egyptians, Greeks, Romans, Mayans, Arabs, and Chinese (to name just a few) all developed their own systems of writing numbers.

Although Roman numerals gained wide currency as the Roman Empire expanded throughout Europe and parts of Asia and Africa, the more advanced system that the Arabs invented turned out to be more useful. Our own number system, the Hindu-Arabic numbers (also called decimal numbers), is closely derived from these early Arabic numbers.

Understanding Number Sequences

Although numbers were invented for counting commodities, as I explain in the preceding section, they were soon put to a wide range of applications. Numbers could be useful for measuring distances, counting money, amassing an army, levying taxes, building pyramids, and lots more.

But beyond their many uses for understanding the external world, numbers also have an internal order all their own. So, numbers are not only an *invention,* but also equally a *discovery:* a landscape that seems to exist independently, with its own structure, mysteries, and even perils.

One path into this new and often strange world is the *number sequence:* an arrangement of numbers according to a rule. In the following sections I introduce you to a variety of number sequences that are useful for making sense of numbers.

Evening the odds

One of the first things you probably heard about numbers is that all of them are either even or odd. For example, you can split an even number of marbles *evenly* into two equal piles. But when you try to divide an odd number of marbles the same way, you always have one *odd,* leftover marble. Here are the first few even numbers:

2   4   6   8   10   12   14   16 ...

You can easily keep the sequence of even numbers going as long as you like. Starting with the number 2, keep adding 2 to get the next number.

Similarly, here are the first few odd numbers:

1   3   5   7   9   11   13   15 ...

The sequence of odd numbers is just as simple to generate. Starting with the number 1, keep adding 2 to get the next number.

Patterns of even or odd numbers are the simplest number patterns around, which is why kids often figure out the difference between even and odd numbers soon after learning to count.

Counting by threes, fours, fives, and so on

After you get used to the concept of counting by numbers greater than one, you can run with it. For example, here's what counting by threes looks like:

3   6   9   12   15   18   21   24 ...

This time, the pattern is generated by starting with 3 and continuing to add 3.

Similarly, here's how to count by fours:

4   8   12   16   20   24   28   32 ...

And here's how to count by fives:

5   10   15   20   25   30   35   40 ...

Counting by a given number is a good way to begin learning the multiplication table for that number, especially for the numbers you're kind of sketchy on. (In general, people seem to have the most trouble multiplying by 7, but 8 and 9 are also unpopular.) In Chapter 3, I show you a few tricks for memorizing the multiplication table once and for all.

These types of sequences are also useful for understanding factors and multiples, which you get a look at in Chapter 8.

Getting square with square numbers

When you study math, sooner or later you probably want to use visual aids to help you see what numbers are telling you. (Later in this book, I show you how one picture can be worth a thousand numbers when I discuss geometry in Chapter 16 and graphing in Chapter 17.)

The tastiest visual aids you'll ever find are those little square cheese-flavored crackers. (You probably have a box sitting somewhere in the pantry. If not, saltine crackers or any other square food works just as well.) Shake a bunch out of a box and place the little squares together to make bigger squares. Figure 1-1 shows the first few.

Figure 1-1: Square numbers.

1	2	3	4	5
6	7	8	9	10
11	12	13	14	15
16	17	18	19	20
21	22	23	24	25

1	2	3	4
5	6	7	8
9	10	11	12
13	14	15	16

1	2	3
4	5	6
7	8	9

1	2
3	4

1

Voila! The square numbers:

$$1 \quad 4 \quad 9 \quad 16 \quad 25 \quad 36 \quad 49 \quad 64 \ldots$$

You get a *square number* by multiplying a number by itself, so knowing the square numbers is another handy way to remember part of the multiplication table. Although you probably remember without help that $2 \times 2 = 4$, you may be sketchy on some of the higher numbers, such as $7 \times 7 = 49$. Knowing the square numbers gives you another way to etch that multiplication table forever into your brain, as I show you in Chapter 3.

Square numbers are also a great first step on the way to understanding exponents, which I introduce later in this chapter and explain in more detail in Chapter 4.

Composing yourself with composite numbers

Some numbers can be placed in rectangular patterns. Mathematicians probably should call numbers like these "rectangular numbers," but instead they chose the term *composite numbers*. For example, 12 is a composite number because you can place 12 objects in rectangles of two different shapes, as shown in Figure 1-2.

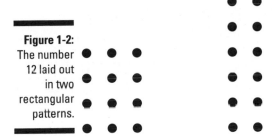

Figure 1-2: The number 12 laid out in two rectangular patterns.

As with square numbers, arranging numbers in visual patterns like this tells you something about how multiplication works. In this case, by counting the sides of both rectangles, you find out the following:

$3 \times 4 = 12$

$2 \times 6 = 12$

Similarly, other numbers such as 8 and 15 can also be arranged in rectangles, as shown in Figure 1-3.

Figure 1-3: Composite numbers, such as 8 and 15, can form rectangles.

As you can see, both of these numbers are quite happy being placed in boxes with at least two rows and two columns. And these visual patterns show this:

$2 \times 4 = 8$

$3 \times 5 = 15$

The word *composite* means that these numbers are *composed of* smaller numbers. For example, the number 15 is composed of 3 and 5 — that is, when you multiply these two smaller numbers, you get 15. Here are all the composite numbers between 1 and 16:

 4   6   8   9   10   12   14   15   16

Notice that all the square numbers (see "Getting square with square numbers") also count as composite numbers because you can arrange them in boxes with at least two rows and two columns. Additionally, lots of other non-square numbers are also composite numbers.

Stepping out of the box with prime numbers

Some numbers are stubborn. Like certain people you may know, these numbers — called *prime numbers* — resist being placed in any sort of a box. Look at how the number 13 is depicted in Figure 1-4, for example.

Figure 1-4:
Unlucky 13,
a prime
example of
a number
that refuses
to fit in a
box.

Try as you may, you just can't make a rectangle out of 13 objects. (That may be one reason the number 13 got a bad reputation as unlucky.) Here are all the prime numbers less than 20:

 2   3   5   7   11   13   17   19

As you can see, the list of prime numbers fills the gaps left by the composite numbers (see the preceding section). Therefore, every counting number is either prime or composite. The only exception is the number 1, which is neither prime nor composite. In Chapter 8, I give you a lot more information about prime numbers and show you how to *decompose* a number — that is, break a composite number down to its prime factors.

Multiplying quickly with exponents

Here's an old question that still causes surprises: Suppose you took a job that paid you just 1 penny the first day, 2 pennies the second day, 4 pennies the third day, and so on, doubling the amount every day, like this:

1	2	4	8	16	32	64	128	256	512 ...

As you can see, in the first ten days of work, you would've earned a little more than $10 (actually, $10.23 — but who's counting?). How much would you earn in 30 days? Your answer may well be, "I wouldn't take a lousy job like that in the first place." At first glance, this looks like a good answer, but here's a glimpse at your second ten days' earnings:

... 1,024   2,048   4,096   8,192   16,384   32,768   65,536
131,072   262,144   524,288 ...

By the end of the second 10 days, your total earnings would be over $10,000. And by the end of the third week, your earnings would top out around $10,000,000! How does this happen? Through the magic of exponents (also called *powers*). Each new number in the sequence is obtained by multiplying the previous number by 2:

$2^1 = 2 = 2$

$2^2 = 2 \times 2 = 4$

$2^3 = 2 \times 2 \times 2 = 8$

$2^4 = 2 \times 2 \times 2 \times 2 = 16$

As you can see, the notation 2^4 means *multiply 2 by itself 4 times.*

You can use exponents on numbers other than 2. Here's another sequence you may be familiar with:

1	10	100	1,000	10,000	100,000	1,000,000...

In this sequence, every number is 10 times greater than the number before it. You can also generate these numbers using exponents:

$10^1 = 10 = 10$

$10^2 = 10 \times 10 = 100$

$10^3 = 10 \times 10 \times 10 = 1,000$

$10^4 = 10 \times 10 \times 10 \times 10 = 10,000$

This sequence is important for defining *place value*, the basis of the decimal number system, which I discuss in Chapter 2. It also shows up when I discuss decimals in Chapter 11 and scientific notation in Chapter 15. You find out more about exponents in Chapter 5.

Looking at the Number Line

As kids outgrow counting on their fingers (and use them only when trying to remember the names of all seven dwarfs), teachers often substitute a picture of the first ten numbers in order, like the one shown in Figure 1-5.

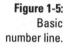 **Figure 1-5:** Basic number line.

This way of organizing numbers is called the *number line*. People often see their first number line — usually made of brightly colored construction paper — pasted above the blackboard in school. The basic number line provides a visual image of the *counting numbers* (also called the *natural numbers*), the numbers greater than 0. You can use it to show how numbers get bigger in one direction and smaller in the other.

In this section, I show you how to use the number line to understand a few basic but important ideas about numbers.

Adding and subtracting on the number line

You can use the number line to demonstrate simple addition and subtraction. These first steps in math become a lot more concrete with a visual aid. Here's the main thing to remember:

✔ As you go *right,* the numbers go *up,* which is *addition* (+).

✔ As you go *left,* the numbers go *down,* which is *subtraction* (–).

For example, 2 + 3 means you *start at* 2 and *jump up* 3 spaces to 5, as illustrated in Figure 1-6.

Figure 1-6:
Moving
through the
number line
from left
to right.

As another example, 6 − 4 means *start at* 6 and *jump down* 4 spaces to 2. That
is, 6 − 4 = 2. See Figure 1-7.

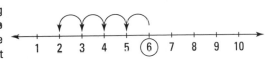

Figure 1-7:
Moving
through the
number line
from right
to left.

You can use these simple up and down rules repeatedly to solve a longer
string of added and subtracted numbers. For example, 3 + 1 − 2 + 4 − 3 − 2
means 3, *up* 1, *down* 2, *up* 4, *down* 3, and *down* 2. In this case, the number line
would show you that 3 + 1 − 2 + 4 − 3 − 2 − 1.

I discuss addition and subtraction in greater detail in Chapter 3.

Getting a handle on nothing, or zero

An important addition to the number line is the number 0, which means
nothing, zilch, nada. Step back a moment and consider the bizarre concept of
nothing. For one thing — as more than one philosopher has pointed out — by
definition, *nothing* doesn't exist! Yet, we routinely label it with the number 0,
as shown in Figure 1-8.

Actually, mathematicians have an even more precise labeling of *nothing* than
zero. It's called the *empty set,* which is sort of the mathematical version of a
box containing nothing. I introduce you to some basic set theory in Chapter 20.

Figure 1-8:
The number
line starting
at 0 and
continuing
with 1, 2,
3, . . . 10.

Nothing sure is a heavy trip to lay on little kids, but they don't seem to mind. They understand quickly that when you have three toy trucks and someone else takes away all three of them, you're left with zero trucks. That is, 3 – 3 = 0. Or, placing this on the number line, 3 – 3 means start at 3 and go down 3, as shown in Figure 1-9.

Figure 1-9:
Starting at 3
and moving
down three.

In Chapter 2, I show you the importance of 0 as a *placeholder* in numbers and discuss how *leading zeros* can be attached to a number without changing its value.

Infinity: Imagining a never-ending story

The arrows at the ends of the number line point onward to a place called *infinity,* which isn't really a place at all, just the idea of *foreverness,* because the numbers go on forever. But what about a million billion trillion quadrillion — do the numbers go even higher than that? The answer is yes, because for any number you name, you can add 1 to it.

The wacky symbol ∞ represents infinity. Remember, though, that ∞ isn't really a number but the *idea* that the numbers go on forever.

Because ∞ isn't a number, you can't technically add the number 1 to it, any more than you can add the number 1 to a cup of coffee or your Aunt Louise. But even if you could, ∞ + 1 would equal ∞.

Taking a negative turn: Negative numbers

When people first find out about subtraction, they often hear that you can't take away more than you have. For example, if you have four pencils, you can take away one, two, three, or even all four of them, but you can't take away more than that.

It isn't long, though, before you find out what any credit card holder knows only too well: You can, indeed, take away more than you have — the result is a *negative number*. For example, if you have $4 and you owe your friend $7, you're $3 in debt. That is, 4 − 7 = −3. The minus sign in front of the 3 means that the number of dollars you have is three less than 0. Figure 1-10 shows how you place negative whole numbers on the number line.

Figure 1-10:
Negative
whole
numbers
on the
number line.

Adding and subtracting on the number line works pretty much the same with negative numbers as with positive numbers. For example, Figure 1-11 shows how to subtract 4 − 7 on the number line.

Figure 1-11:
Subtracting
4 − 7 on the
number line.

You find out all about working with negative numbers in Chapter 4.

Placing 0 and the negative counting numbers on the number line expands the set of counting numbers to the set of *integers*. I discuss the integers in further detail later in this chapter.

Multiplying the possibilities

Suppose you start at 0 and circle every other number on a number line, as shown in Figure 1-12. As you can see, all the even numbers are now circled. In other words, you've circled all the *multiples of two*. (You can find out more about multiples in Chapter 8.) You can now use this number line to multiply any number by two. For example, suppose you want to multiply 5×2. Just start at 0 and jump 5 circled spaces to the right.

Figure 1-12:
Multiplying
5×2 using
the number
line.

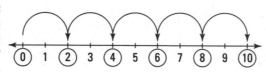

This number line shows you that $5 \times 2 = 10$.

Similarly, to multiply -3×2, start at 0 and jump 3 circled spaces to the left (that is, in the negative direction). Figure 1-13 shows you that $-3 \times 2 = -6$. What's more, you can now see why multiplying a negative number by a positive number always gives you a negative result. (I talk about multiplying by negative numbers in Chapter 4.)

Figure 1-13:
$-3 \times 2 = -6$,
as depicted
on the
number line.

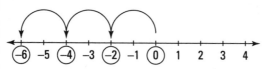

Multiplying on the number line works no matter what number you count off by. For example, in Figure 1-14 I count off by 5s.

Figure 1-14:
Number line
counted off
by 5s.

This time, the only numbers I've marked are the *multiples of 5,* so I can use this number line to multiply any number by 5. For example, Figure 1-15 shows how to multiply 2×5.

Figure 1-15:
Multiplying
2×5 with
the number
line.

So $2 \times 5 = 10$, the same result as when you multiply 5×2. This result is an example of the *commutative property of multiplication* — you can reverse the order of any multiplication problem and still get the same answer. (I discuss the commutative property in Chapter 4.)

Dividing things up

You can also use the number line to divide. For example, suppose you want to divide 6 by some other number. First, draw a number line that begins at 0 and ends at 6, as in Figure 1-16.

Figure 1-16:
Number line
from 0 to 6.

Now, to find the answer to $6 \div 2$, just split this number line into two equal parts, as shown in Figure 1-17. This split (or *division*) occurs at 3, showing you that $6 \div 2 = 3$.

Figure 1-17:
Getting the
answer to
$6 \div 2$ by
splitting the
number line.

Similarly, to divide $6 \div 3$, split the same number line into three equal parts, as in Figure 1-18. This time you have two splits, so use the one closest to 0. This number line shows you that $6 \div 3 = 2$.

Figure 1-18:
Dividing
$6 \div 3$ with
the number
line.

But suppose you want to use the number line to divide a small number by a larger number. For example, maybe you want to know the answer to $3 \div 4$. Following the method I show you earlier, first draw a number line from 0 to 3. Then split it into four equal parts. Unfortunately, none of these splits has landed on a number. That's not a mistake. You just have to add some new numbers to the number line, as you can see in Figure 1-19.

Figure 1-19:
Fractions on
the number
line.

Welcome to the world of *fractions*. With the number line labeled properly, you can see that the split closest to 0 is ¾. This image tells you that $3 \div 4 = ¾$.

The similarity of the expression $3 \div 4$ and the fraction ¾ is no accident. Division and fractions are closely related. When you divide, you cut things up into equal parts, and fractions are often the result of this process. (I explain the connection between division and fractions in more detail in Chapters 9 and 10.)

Discovering the space in between: Fractions

Fractions help you fill in a lot of the spaces on the number line that fall between the counting numbers. For example, Figure 1-20 shows a close-up of a number line from 0 to 1.

Figure 1-20:
Number line
depicting
some
fractions
from 0 to 1.

This number line may remind you of a ruler or a tape measure, with lots of tiny fractions filled in. And in fact, rulers and tape measures really are portable number lines that allow carpenters, engineers, and savvy do-it-yourselfers to measure the length of objects with precision.

The addition of fractions to the number line expands the set of integers to the set of *rational numbers*. I discuss the rational numbers in greater detail in Chapter 25.

In fact, no matter how small things get in the real world, you can always find a tiny fraction to approximate it as closely as you need. Between any two fractions on the number line, you can always find another fraction. Mathematicians call this trait the *density* of fractions on the real number line, and this type of density is a topic in a very advanced area of math called *real analysis*.

Four Important Sets of Numbers

In the preceding section, you see how the number line grows in both the positive and negative directions and fills in with lots of numbers in between. In this section, I provide a quick tour of how numbers fit together as a set of nested systems, one inside the other.

When I talk about a *set* of numbers, I'm really just talking about a group of numbers. You can use the number line to deal with four important sets of numbers:

- ✔ **Counting numbers (also called natural numbers):** The set of numbers beginning 1, 2, 3, 4 ... and going on infinitely

- ✔ **Integers:** The set of counting numbers, zero, and negative counting numbers

- ✔ **Rational numbers:** The set of integers and fractions

- ✔ **Real numbers:** The set of rational and irrational numbers

The sets of counting numbers, integers, rational, and real numbers are nested, one inside another. This nesting of one set inside another is similar to the way that a city (for example, Boston) is inside a state (Massachusetts), which is inside a country (the United States), which is inside a continent (North America). The set of counting numbers is inside the set of integers, which is inside the set of rational numbers, which is inside the set of real numbers.

Counting on the counting numbers

The set of *counting numbers* is the set of numbers you first count with, starting with 1. Because they seem to arise naturally from observing the world, they're also called the *natural numbers:*

$$1 \quad 2 \quad 3 \quad 4 \quad 5 \quad 6 \quad 7 \quad 8 \quad 9 \dots$$

The counting numbers are infinite, which means they go on forever.

When you add two counting numbers, the answer is always another counting number. Similarly, when you multiply two counting numbers, the answer is always a counting number. Another way of saying this is that the set of counting numbers is *closed* under both addition and multiplication.

Introducing integers

The set of *integers* arises when you try to subtract a larger number from a smaller one. For example, $4 - 6 = -2$. The set of integers includes the following:

- ✔ The counting numbers
- ✔ Zero
- ✔ The negative counting numbers

Here's a partial list of the integers:

$$\dots -4 \quad -3 \quad -2 \quad -1 \quad 0 \quad 1 \quad 2 \quad 3 \quad 4 \dots$$

Like the counting numbers, the integers are closed under addition and multiplication. Similarly, when you subtract one integer from another, the answer is always an integer. That is, the integers are also closed under subtraction.

Staying rational

Here's the set of *rational numbers*:

- ✔ Integers
  - Counting numbers
  - Zero
  - Negative counting numbers
- ✔ Fractions

Like the integers, the rational numbers are closed under addition, subtraction, and multiplication. Furthermore, when you divide one rational number by another, the answer is always a rational number. Another way to say this is that the rational numbers are closed under division.

Getting real

Even if you filled in all the rational numbers, you'd still have points left unlabeled on the number line. These points are the irrational numbers.

An *irrational number* is a number that's neither a whole number nor a fraction. In fact, an irrational number can only be approximated as a *non-repeating decimal*. In other words, no matter how many decimal places you write down, you can always write down more; furthermore, the digits in this decimal never become repetitive or fall into any pattern. (For more on repeating decimals, see Chapter 11.)

The most famous irrational number is π (you find out more about π when I discuss the geometry of circles in Chapter 17.):

π = 3.14159265358979323846264338327950288419716939937510 ...

Together, the rational and irrational numbers make up the *real numbers*, which comprise every point on the number line. In this book, I don't spend too much time on irrational numbers, but just remember that they're there for future reference.

Chapter 2

It's All in the Fingers — Numbers and Digits

When you're counting, *ten* seems to be a natural stopping point — a nice, round number. The fact that our ten fingers are matched up so nicely with numbers may seem like a happy accident. But of course, it's no accident at all. Fingers were the first calculator that humans possessed. Our number system — Hindu-Arabic numbers — is based on the number ten because humans have ten fingers rather than eight or twelve. In fact, the very word *digit* has two meanings: numerical symbol and finger.

In this chapter, I show you how place value turns digits into numbers. I also show you when 0 is an important placeholder in a number and why leading zeros don't change the value of a number. And I show you how to read and write long numbers. After that, I discuss two important skills: rounding numbers and estimating values. I also note when estimating is likely to result in a misleading answer and show you how to avoid bad estimates and make better ones.

Telling the difference between numbers and digits

Sometimes, people confuse numbers and digits. For the record, here's the difference:

- A *digit* is a single numerical symbol, from 0 to 9.

- A *number* is a string of one or more digits.

For example, 7 is both a digit and a number. In fact, it's a one-digit number. However, 15 is a string of two digits, so it's a number — a two-digit number. And 426 is a three-digit number. You get the idea.

In a sense, a digit is like a letter of the alphabet. By themselves, the uses of 26 letters, *A* through *Z,* are limited. (How much can you do with a single letter such as *K* or *W?*) Only when you begin using strings of letters as building blocks to spell words does the power of letters become apparent. Similarly, the ten digits, 0 through 9, have a limited usefulness until you begin building strings of digits — that is, numbers.

Knowing Your Place Value

The number system you're most familiar with — Hindu-Arabic numbers — has ten familiar digits:

$$0 \quad 1 \quad 2 \quad 3 \quad 4 \quad 5 \quad 6 \quad 7 \quad 8 \quad 9$$

Yet with only ten digits, you can express numbers as high as you care to go. In this section, I show you how it happens.

Counting to ten and beyond

The ten digits in our number system allow you to count from 0 to 9. All higher numbers are produced using place value. *Place value* assigns a digit a greater or lesser value depending on where it appears in a number. Each place in a number is ten times greater than the place to its immediate right.

To understand how a whole number gets its value, suppose you write the number 45,019 all the way to the right in Table 2-1, one digit per cell, and add up the numbers you get.

Table 2-1	45,019 Displayed in a Place-Value Chart							
Millions			**Thousands**			**Ones**		
Hundred Millions	Ten Millions	Mil-lions	Hundred Thousands	Ten Thou-sands	Thou-sands	Hun-dreds	Tens	Ones
				4	5	0	1	9

You have 4 ten thousands, 5 thousands, 0 hundreds, 1 ten, and 9 ones. The chart shows you that the number breaks down as follows:

$$45,019 = 40,000 + 5,000 + 0 + 10 + 9$$

In this example, notice that the presence of the digit 0 in the hundreds place means that *zero* hundreds are added to the number.

Telling placeholders from leading zeros

Although the digit 0 adds no value to a number, it acts as a *placeholder* to keep the other digits in their proper places. For example, the number 5,001,000 breaks down into 5,000,000 + 1,000. Suppose, however, you decide to leave all the 0s out of the chart. Table 2-2 shows what you'd get.

Table 2-2	5,001,000 Displayed *Incorrectly* without Place-Holding Zeros							
Millions			**Thousands**			**Ones**		
Hundred Millions	Ten Millions	Mil-lions	Hundred Thousands	Ten Thou-sands	Thou-sands	Hun-dreds	Tens	Ones
							5	1

The chart tells you that 5,001,000 = 50 + 1. Clearly, this answer is wrong!

As a rule, when a 0 appears to the right of *at least one digit other than 0,* it's a placeholder. Place-holding zeros are important — always include them when you write a number. However, when a 0 appears to the left of every digit other than 0, it's a *leading zero.* Leading zeros serve no purpose in a number, so dropping them is customary. For example, place the number 003,040,070 on the chart (see Table 2-3).

Table 2-3			3,040,070 Displayed with Two Leading Zeros					
Millions			*Thousands*			*Ones*		
Hundred Millions	Ten Millions	Mil-lions	Hundred Thousands	Ten Thou-sands	Thou-sands	Hun-dreds	Tens	Ones
0	0	3	0	4	0	0	7	0

The first two 0s in the number are leading zeros because they appear to the left of the 3. You can drop these 0s from the number, leaving you with 3,040,070. The remaining 0s are all to the right of the 3, so they're placeholders — be sure to write them in.

Reading long numbers

When you write a long number, you use commas to separate periods. *Periods* are simply groups of three numbers. They help make long numbers more readable. For example, here's about as long a number as you'll ever see:

234,845,021,349,230,467,304

Table 2-4 shows a larger version of the place-value chart.

Table 2-4	A Place-Value Chart Separated into Periods					
Quintillions	**Quadrillions**	**Trillions**	**Billions**	**Millions**	**Thousands**	**Ones**
234	845	021	349	230	467	304

This version of the chart helps you read the number. Begin all the way to the left and read, "Two hundred thirty-four quintillion, eight hundred forty-five quadrillion, twenty-one trillion, three hundred forty-nine billion, two hundred thirty million, four hundred sixty-seven thousand, three hundred four."

When you read and write whole numbers, don't say the word *and*. In math, the word *and* means you have a decimal point. That's why when you write a check, you save the word *and* for the number of cents, which is expressed as a decimal or fraction. (I discuss decimals in Chapter 11.)

Close Enough for Rock 'n' Roll: Rounding and Estimating

As numbers get longer, calculations become tedious, and you're more likely to make a mistake or just give up. When you're working with long numbers, simplifying your work by rounding numbers and estimating values is sometimes helpful. At other times, you may want to round off an answer after you do your calculations.

When you *round* a number, you change some of its digits to place-holding zeros. Rounding numbers allows you to change a difficult number to an easier one. And when you *estimate* a value, you work with rounded numbers to find an approximate answer to a problem. In this section, you build both skills. I also show you when you're likely to make a bad estimate and how to avoid it.

Rounding numbers

Rounding numbers makes long numbers easier to work with. In this section, I show you how to round numbers to the nearest ten, hundred, thousand, and beyond.

Rounding numbers to the nearest ten

The simplest kind of rounding you can do is with two-digit numbers. When you round a two-digit number to the nearest ten, you simply bring it up or down to the *nearest* number that ends in 0. For example,

$$39 \rightarrow 40 \qquad 51 \rightarrow 50 \qquad 73 \rightarrow 70$$

Even though numbers ending in 5 are in the middle, always round them up to the next-highest number that ends in 0:

$$15 \rightarrow 20 \qquad 35 \rightarrow 40 \qquad 85 \rightarrow 90$$

Numbers in upper 90s get rounded up to 100:

$$99 \rightarrow 100 \qquad 95 \rightarrow 100 \qquad 94 \rightarrow 90$$

After you know how to round a two-digit number, you can round just about any number. For example, to round most longer numbers to the nearest ten, just focus on the ones and tens digits:

$$734 \rightarrow 730 \qquad 1{,}488 \rightarrow 1{,}490 \qquad 12{,}345 \rightarrow 12{,}350$$

Occasionally, a small change to the ones and tens digits affects the other digits. (This is a lot like when the odometer in your car rolls a bunch of 9s over to 0s.) For example:

$$899 \rightarrow 900 \qquad 1{,}097 \rightarrow 1{,}100 \qquad 9{,}995 \rightarrow 10{,}000$$

Rounding numbers to the nearest hundred and beyond

To round numbers to the nearest hundred, or thousand, or beyond, focus only on two digits: the digit in the place you're rounding to and the digit to its immediate right. Change all other digits to the right of these two digits to 0s. For example, suppose you want to round 642 to the nearest hundred. Focus on the hundreds digit (6) and the digit to its immediate right (4):

642

I've underlined these two digits. Now, just round these two digits as if you were rounding to the nearest ten, and change the digit to the right of them to a 0:

$$642 \rightarrow 600$$

Here are a few more examples of rounding numbers to the nearest hundred:

$$7{,}891 \rightarrow 7{,}900 \qquad 15{,}753 \rightarrow 15{,}800 \qquad 99{,}961 \rightarrow 100{,}000$$

When rounding numbers to the nearest thousand, underline the thousands digit and the digit to its immediate right. Round the number by focusing only on the two underlined digits and, when you're done, change all digits to the right of these to 0s:

$$4{,}984 \rightarrow 5{,}000 \qquad 78{,}521 \rightarrow 79{,}000 \qquad 1{,}099{,}304 \rightarrow 1{,}099{,}000$$

Even when rounding to the nearest million, the same rules apply:

$$1{,}234{,}567 \rightarrow 1{,}000{,}000 \qquad 78{,}883{,}958 \rightarrow 79{,}000{,}000$$

Estimating value to make problems easier

After you know how to round numbers, you can use this skill in estimating values. Estimating saves you time by allowing you to avoid complicated computations and still get an approximate answer to a problem.

When you get an approximate answer, you don't use an equal sign; instead, you use this wavy symbol, which means *is approximately equal to:* ≈.

Suppose you want to add these numbers: 722 + 506 + 383 + 1,279 + 91 + 811. This computation is tedious, and you may make a mistake. But you can make the addition easier by first rounding all the numbers to the nearest hundred and then adding:

$$\approx 700 + 500 + 400 + 1300 + 100 + 800 = 3,800$$

The approximate answer is 3,800. This answer isn't very far off from the exact answer, which is 3,792.

Similarly, suppose you want to multiply 879 × 618. Again, this computation doesn't look like a lot of fun. Instead, you can round these numbers to the nearest hundred:

$$\approx 900 \times 600 = 540,000$$

This time, the approximate answer is 540,000, and the exact answer is 543,222. Again, not a bad estimate.

Estimating is useful, but it can also lead to results that aren't very close to the right answer. You're most likely to arrive at a bad estimate when you

- ✔ Round numbers that are close to the middle of the range
- ✔ Round too many numbers in the same direction (either up or down)
- ✔ Multiply or divide rounded numbers

For example, suppose you want to multiply 349 × 243. You start by rounding both of these numbers:

$$\approx 300 \times 200 = 60,000$$

Your estimate is 60,000, but the actual answer is 84,807 — not a very good approximation. What happened? First, notice that 349 is very close to the middle of the range between 300 and 400. And 247 is close to the middle of the range between 200 and 300. So when you rounded these numbers, you changed their values a lot. Second, you rounded both numbers *down* to lower values. That's why the estimate ended up so much lower than the real answer. Third, you multiplied. In general, multiplying and dividing throw off estimates more than addition and subtraction do.

Chapter 3

The Big Four: Addition, Subtraction, Multiplication, and Division

*W*hen most folks think of math, the first thing that comes to mind is four little (or not-so-little) words: addition, subtraction, multiplication, and division. I call these operations the *Big Four* all through the book.

In this chapter, I introduce you (or reintroduce you) to these little gems. Although I assume you're already familiar with the Big Four, this chapter reviews these operations, taking you from what you may have missed to what you need to succeed as you move onward and upward in math.

Adding Things Up

Addition is the first operation you find out about, and it's almost everybody's favorite. It's simple, friendly, and straightforward. No matter how much you worry about math, you probably never lost a minute of sleep over addition. Addition is all about bringing things together, which is a positive thing. For example, suppose you and I are standing in line to buy tickets for a movie. I have $15 and you have only $5. I could lord it over you and make you feel crummy that I can go to the movies and you can't. Or instead, you and I can join forces, adding together my $15 and your $5 to make $20. Now, not only can we both see the movie, but we may even be able to buy some popcorn, too.

Addition uses only one sign — the plus sign (+): Your equation may read 2 + 3 = 5 or 12 + 2 = 14 or 27 + 44 = 71, but the plus sign always means the same thing.

When you add two numbers together, those two numbers are called *addends,* and the result is called the *sum.* So in the first example, the addends are 2 and 3, and the sum is 5.

In line: Adding larger numbers in columns

When you want to add larger numbers, stack them on top of each other so that the ones digits line up in a column, the tens digits line up in another column, and so on. (Chapter 1 has the scoop on digits and place value.) Then add column by column, starting from the ones column on the right. Not surprisingly, this method is called *column addition.* Here's how you add 55 + 31 + 12. First add the ones column:

$$
\begin{array}{r}
55 \\
31 \\
+\ 12 \\
\hline
8
\end{array}
$$

Next, move to the tens column:

$$
\begin{array}{r}
55 \\
31 \\
+\ 12 \\
\hline
98
\end{array}
$$

This problem shows you that 55 + 31 + 12 = 98.

Carry on: Dealing with two-digit answers

Sometimes when you're adding a column, the sum is a two-digit number. In that case, you need to write down the ones digit of that number and *carry* the tens digit over to the next column to the left — that is, write this digit above the column so you can add it with the rest of the numbers in that column. For example, suppose you want to add 376 + 49 + 18. In the ones column, 6 + 9 + 8 = 23, so write down the 3 and carry the 2 over to the top of the tens column:

$$
\begin{array}{r}
2 \\
376 \\
49 \\
+\ 18 \\
\hline
3 \\
\end{array}
$$

Now continue by adding the tens column. In this column, 2 + 7 + 4 + 1 = 14, so write down the 4 and carry the 1 over to the top of the hundreds column:

$$
\begin{array}{r}
12 \\
376 \\
49 \\
+\ 18 \\
\hline
43 \\
\end{array}
$$

Continue adding in the hundreds column:

$$
\begin{array}{r}
12 \\
376 \\
49 \\
+\ 18 \\
\hline
443 \\
\end{array}
$$

This problem shows you that 376 + 49 + 18 = 443.

Take It Away: Subtracting

Subtraction is usually the second operation you discover, and it's not much harder than addition. Still, there's something negative about subtraction — it's all about who has more and who has less. Suppose you and I have been running on treadmills at the gym. I'm happy because I ran 3 miles, but then you start bragging that you ran 10 miles. You subtract and tell me that I should be very impressed that you ran 7 miles farther than I did. (But with an attitude like that, don't be surprised if you come back from the showers to find your running shoes filled with liquid soap!)

As with addition, subtraction has only one sign: the minus sign (–). You end up with equations such as 4 – 1 = 3 and 14 – 13 = 1 and 93 – 74 = 19.

When you subtract one number from another, the result is called the *difference*. This term makes sense when you think about it: When you subtract, you find out the difference between a higher number and a lower one.

In subtraction, the first number is called the *subtrahend,* and the second number is called the *minuend.* But almost nobody ever remembers which is which, so when I talk about subtraction, I prefer to say *the first number* and *the second number.*

One of the first things you probably heard about subtraction is that you can't take away more than you start with. In that case, the second number can't be larger than the first. And if the two numbers are the same, the result is always 0. For example, $3 - 3 = 0$ and $11 - 11 = 0$ and $1776 - 1776 = 0$. Later, someone breaks the news that you *can* take away more than you have. When you do, though, you need to place a minus sign in front of the difference to show that you have a *negative number,* a number below 0:

$$4 - 5 = -1$$
$$10 - 13 = -3$$
$$88 - 99 = -11$$

When subtracting a larger number from a smaller number, remember the words *switch* and *negate.* That is, *switch* the order of the two numbers and do the subtraction as you normally would, but at the end *negate* the result by attaching a minus sign. For example, to find $10 - 13$, you switch the order of these two numbers, giving you $13 - 10$, which equals 3; then *negate* this result to get -3. That's why $10 - 13 = -3$.

The minus sign gets double duty, so don't get confused. When you stick a minus sign *between* two numbers, it means *the first number minus the second number.* But when you attach it to the *front* of a number, it means that this number is a negative number.

Flip to Chapter 1 to see how negative numbers work on the number line. I also go into more detail on negative numbers and the Big Four operations in Chapter 4.

Columns and stacks: Subtracting larger numbers

To subtract larger numbers, stack one on top of the other as you do with addition. (For subtraction, however, don't stack more than two numbers — put the larger number on top and the smaller under it.) For example, suppose you want to subtract 386 – 54. To start out, stack the two numbers and begin subtracting in the ones column: 6 – 4 = 2:

$$
\begin{array}{r}
386 \\
-\ 54 \\
\hline
2
\end{array}
$$

Next, move to the tens column and subtract 8 – 5 to get 3:

$$
\begin{array}{r}
386 \\
-\ 54 \\
\hline
32
\end{array}
$$

Finally, move to the hundreds column. This time, 3 – 0 = 3:

$$
\begin{array}{r}
386 \\
-\ 54 \\
\hline
332
\end{array}
$$

This problem shows you that 386 – 54 = 332.

Can you spare a ten? Borrowing to subtract

Sometimes, the top digit in a column is smaller than the bottom digit in that column. In that case, you need to *borrow* from the next column to the left. Borrowing is a two-step process:

1. **Subtract 1 from the top number in the column directly to the left.**

   Cross out the number you're borrowing from, subtract 1, and write the answer above the number you crossed out.

2. **Add 10 to the top number in the column you were working in.**

For example, suppose you want to subtract 386 – 94. The first step is to subtract 4 from 6 in the ones column, which gives you 2:

$$
\begin{array}{r}
386 \\
-\ 94 \\
\hline
2
\end{array}
$$

When you move to the tens column, however, you find that you need to subtract 8 – 9. Because 8 is smaller than 9, you need to borrow from the hundreds column. First, cross out the 3 and replace it with a 2, because 3 – 1 = 2:

$$\begin{array}{r} 2 \\ \cancel{3}86 \\ -\ 94 \\ \hline 2 \end{array}$$

Next, place a 1 in front of the 8, changing it to an 18, because 8 + 10 = 18:

$$\begin{array}{r} 2 \\ \cancel{3}\ 18\ 6 \\ -\ \ 9\ 4 \\ \hline 2 \end{array}$$

Now you can subtract in the tens column: 18 – 9 = 9:

$$\begin{array}{r} 2\ 18\ 6 \\ -\ \ 9\ 4 \\ \hline 9\ 2 \end{array}$$

The final step is simple: 2 – 0 = 2:

$$\begin{array}{r} 2\ 18\ 6 \\ -\ \ 9\ 4 \\ \hline 2\ \ 9\ 2 \end{array}$$

Therefore, 386 – 94 = 292.

In some cases, the column directly to the left may not have anything to lend. Suppose, for instance, you want to subtract 1,002 – 398. Beginning in the ones column, you find that you need to subtract 2 – 8. Because 2 is smaller than 8, you need to borrow from the next column to the left. But the digit in the tens column is a 0, so you can't borrow from there because the cupboard is bare, so to speak:

$$\begin{array}{r} 1,002 \\ -\ 398 \\ \hline \end{array}$$

When borrowing from the next column isn't an option, you need to borrow from the *nearest non-zero column to the left*.

In this example, the column you need to borrow from is the thousands column. First, cross out the 1 and replace it with a 0. Then place a 1 in front of the 0 in the hundreds column:

```
  0
  1 10 0 2
  – 3 9 8
```

Now, cross out the 10 and replace it with a 9. Place a 1 in front of the 0 in the tens column:

```
  0  9
  1 10 10 2
  – 3  9 8
```

Finally, cross out the 10 in the tens column and replace it with a 9. Then place a 1 in front of the 2:

```
  0  9  9
  1 10 10 12
  – 3  9  8
```

At last, you can begin subtracting in the ones column: $12 – 8 = 4$:

```
  0  9  9
  1 10 10 12
  – 3  9  8
              4
```

Then subtract in the tens column: $9 – 9 = 0$:

```
  0  9  9
  1 10 10 12
  – 3  9  8
        0  4
```

Then subtract in the hundreds column: $9 – 3 = 9$:

```
  0  9  9
  1 10 10 12
  – 3  9  8
     6  0  4
```

Because nothing is left in the thousands column, you don't need to subtract anything else. Therefore, $1,002 – 398 = 604$.

Multiplying

Multiplication is often described as a sort of shorthand for repeated addition. For example,

4×3 means *add 4 to itself 3 times:* $4 + 4 + 4 = 12$

9×6 means *add 9 to itself 6 times:* $9 + 9 + 9 + 9 + 9 + 9 = 54$

100×2 means *add 100 to itself 2 times:* $100 + 100 = 200$

Although multiplication isn't as warm and fuzzy as addition, it's a great timesaver. For example, suppose you coach a Little League baseball team and you've just won a game against the toughest team in the league. As a reward, you promised to buy three hot dogs for each of the nine players on the team. To find out how many hot dogs you need, you could add 3 together 9 times. Or you can save time by multiplying 3 times 9, which gives you 27. Therefore, you need 27 hot dogs (plus a whole lot of mustard and sauerkraut).

When you multiply two numbers, the two numbers that you're multiplying are called *factors,* and the result is the *product.*

In multiplication, the first number is also called the *multiplicand* and the second number is the *multiplier.* But almost nobody ever remembers these words.

Signs of the times

When you're first introduced to multiplication, you use the *times sign* ($\times$). However, algebra uses the letter *x* a lot, which looks similar to the times sign, so people often choose to use other multiplication symbols for clarity. As you move onwards and upwards on your math journey, you should be aware of the conventions I discuss in the following sections.

Arriving on the dot

In math beyond arithmetic, the symbol · replaces $\times$. For example,

$4 \cdot 2 = 8$ means $4 \times 2 = 8$

$6 \cdot 7 = 42$ means $6 \times 7 = 42$

$53 \cdot 11 = 583$ means $53 \times 11 = 583$

That's all there is to it: Just use the · symbol anywhere you would've used the standard times sign (×).

Speaking parenthetically

In math beyond arithmetic, using parentheses *without* another operator stands for multiplication. The parentheses can enclose the first number, the second number, or both numbers. For example,

$$3(5) = 15 \qquad \text{means} \quad 3 \times 5 = 15$$
$$(8)7 = 56 \qquad \text{means} \quad 8 \times 7 = 56$$
$$(9)(10) = 90 \quad \text{means} \quad 9 \times 10 = 90$$

However, notice that when you place another operator between a number and a parenthesis, that operator takes over. For example,

$$3 + (5) = 8 \qquad \text{means} \quad 3 + 5 = 8$$
$$(8) - 7 = 1 \qquad \text{means} \quad 8 - 7 = 1$$
$$(9) \cdot (10) = 90 \quad \text{means} \quad 9 \times 10 = 90$$

Note: In the third example, you don't really need the ·, but it isn't doing any harm, either.

Memorizing the multiplication table

You may consider yourself among the multiplicationally challenged. That is, you consider being called upon to remember 9×7 a tad less appealing than being dropped from an airplane while clutching a parachute purchased from the trunk of some guy's car. If so, then this section is for you.

Looking at the old multiplication table

One glance at the old multiplication table, Table 3-1, reveals the problem. If you saw the movie *Amadeus,* you may recall that Mozart was criticized for writing music that had "too many notes." Well, in my humble opinion, the multiplication table has too many numbers.

Table 3-1			The Monstrous Standard Multiplication Table						
	0	*1*	*2*	*3*	*4*	*5*	*6*	*7*	*8*	*9*
0	0	0	0	0	0	0	0	0	0	0
1	0	1	2	3	4	5	6	7	8	9
2	0	2	4	6	8	10	12	14	16	18
3	0	3	6	9	12	15	18	21	24	27
4	0	4	8	12	16	20	24	28	32	36
5	0	5	10	15	20	25	30	35	40	45
6	0	6	12	18	24	30	36	42	48	54
7	0	7	14	21	28	35	42	49	56	63
8	0	8	16	24	32	40	48	56	64	72
9	0	9	18	27	36	45	54	63	72	81

I don't like the multiplication table any more than you do. Just looking at it makes my eyes glaze over. With 100 numbers to memorize, no wonder so many folks just give up and carry a calculator.

Introducing the short multiplication table

If the multiplication table from Table 3-1 were smaller and a little more manageable, I'd like it a lot more. So here's my short multiplication table, in Table 3-2.

Table 3-2		The Short Multiplication Table					
		3	*4*	*5*	*6*	*7*	*8*	*9*
3		9	12	15	18	21	24	27
4			16	20	24	28	32	36
5				25	30	35	40	45
6					36	42	48	54
7						49	56	63
8							64	72
9								81

As you can see, I've gotten rid of a bunch of numbers. In fact, I've reduced the table from 100 numbers to 28. I've also shaded 11 of the numbers I've kept.

Is just slashing and burning the sacred multiplication table wise? Is it even legal? Well, of course it is! After all, the table is just a tool, like a hammer. If a hammer's too heavy to pick up, then you ought to buy a lighter one. Similarly, if the multiplication table is too big to work with, then you need a smaller one. Besides, I've removed only the numbers you don't need. For example, the condensed table doesn't include rows or columns for 0, 1, or 2. Here's why:

- ✔ Any number multiplied by 0 is 0 (people call this trait the *zero property of multiplication*).

- ✔ Any number multiplied by 1 is that number itself (which is why mathematicians call 1 the *multiplicative identity*).

- ✔ Multiplying by 2 is fairly easy; if you can count by 2s — 2, 4, 6, 8, 10, and so forth — you can multiply by 2.

The rest of the numbers I've gotten rid of are redundant. (And not just redundant, but also repeated, extraneous, and unnecessary!) For example, any way you slice it, 3×5 and 5×3 are both 15 (you can switch the order of the factors because multiplication is *commutative* — see Chapter 4 for details). In my condensed table, I've simply removed the clutter.

So what's left? Just the numbers you need. These numbers include a gray row and a gray diagonal. The gray row is the *5 times table,* which you probably know pretty well. (In fact, they may evoke a childhood memory of running to find a hiding place on a warm spring day while one of your friends counted in a loud voice: 5, 10, 15, 20, ...)

The numbers on the gray diagonal are the square numbers. As I discuss in Chapter 1, when you multiply any number by itself, the result is a *square number.* You probably know these numbers better than you think.

Getting to know the short multiplication table

In about an hour, you can make huge strides in memorizing the multiplication table. To start out, make a set of flash cards that give a multiplication problem on the front and the answer on the back. They may look like Figure 3-1.

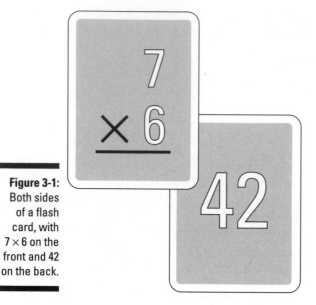

Figure 3-1:
Both sides
of a flash
card, with
7×6 on the
front and 42
on the back.

Remember, you need to make only 28 flash cards — one for every example in Table 3-2. Split these 28 into two piles — a "gray" pile with 11 cards and a "white" pile with 17. (You don't have to color the cards gray and white; just keep track of which pile is which according to the colors in Table 3-2.) Then begin:

1. **5 minutes:** Work with the gray pile, going through it one card at a time. If you get the answer right, put that card on the bottom of the pile. If you get it wrong, put it in the middle so you get another chance at it more quickly.

2. **10 minutes:** Switch to the white pile and work with it in the same way.

3. **15 minutes:** Repeat Steps 1 and 2.

Now take a break. Really — the break is important to rest your brain. Come back later in the day and do the same thing.

After you're done with this exercise, you should find going through all 28 cards with almost no mistakes to be fairly easy. At this point, feel free to make cards for the rest of the standard times table — you know, the cards with all the 0, 1, 2 times tables on them and the redundant problems — mix all 100 cards together, and amaze your family and friends.

TIP

To the nines: A slick trick

Here's a trick to help you remember the 9 times table. To multiply any one-digit number by 9,

1. **Subtract 1 from the number and jot down the answer.**

   For example, suppose you want to multiply 7×9. Here, $7 - 1 = 6$.

2. **Jot down a second number so that, together, the two numbers you wrote add up to 9. You've just written the answer you were looking for.**

   Adding, you get $6 + 3 = 9$. So $7 \times 9 = 63$.

As another example, suppose you want to multiply 8×9:

$$8 - 1 = 7$$

$$7 + 2 = 9$$

So $8 \times 9 = 72$. This trick works for every one-digit number except 0 (but you already know that $0 \times 9 = 0$).

Double digits: Multiplying larger numbers

The main reason to know the multiplication table is so you can more easily multiply larger numbers. For example, suppose you want to multiply 53×7. Start by stacking these numbers one on top of another with a line underneath, and then multiply 3 by 7. Because $3 \times 7 = 21$, write down the 1 and carry the 2:

$$
\begin{array}{r}
2 \\
53 \\
\times\ 7 \\
\hline
1
\end{array}
$$

Next, multiply 7 by 5. This time, 5×7 is 35. But you also need to add the 2 that you carried over, which makes the result 37. Because 5 and 7 are the last numbers to multiply, you don't have to carry, so write down the 37 — you find that $53 \times 7 = 371$:

$$
\begin{array}{r}
2 \\
53 \\
\times\ 7 \\
\hline
371
\end{array}
$$

When multiplying larger numbers, the idea is similar. For example, suppose you want to multiply 53 by 47. (The first few steps — multiplying by the

7 in 47 — are the same, so I pick up with the next step.) Now you're ready to multiply by the 4 in 47. But remember that this 4 is in the *tens* column, so it really means 40. So to begin, put a 0 directly under the 1 in 371:

$$
\begin{array}{r}
53 \\
\times\,47 \\
\hline
371 \\
\mathbf{0}
\end{array}
$$

This 0 acts as a placeholder so that this row is arranged properly. (See Chapter 1 for more about placeholding zeros.)

When multiplying by larger numbers with two digits or more, use one placeholding zero when multiplying by the tens digit, two placeholding zeros when multiplying the hundreds digit, three zeros when multiplying by the thousands digit, and so forth.

Now you multiply 3×4 to get 12, so write down the 2 and carry the 1:

$$
\begin{array}{r}
\mathbf{1} \\
53 \\
\times\,47 \\
\hline
371 \\
\mathbf{20}
\end{array}
$$

Continuing, multiply 5×4 to get 20 and then add the 1 that you carried over, giving a result of 21:

$$
\begin{array}{r}
\mathbf{1} \\
53 \\
\times\,47 \\
\hline
371 \\
\mathbf{21}20
\end{array}
$$

To finish up, add the two products (the multiplication results):

$$
\begin{array}{r}
53 \\
\times\,47 \\
\hline
371 \\
+\;2120 \\
\hline
\mathbf{2{,}491}
\end{array}
$$

So $53 \times 47 = 2{,}491$.

Doing Division Lickety-Split

The last of the Big Four operations is *division*. Division literally means splitting things up. For example, suppose you're a parent on a picnic with your three children. You've brought along 12 pretzel sticks as snacks and want to split them fairly so that each child gets the same number (don't want to cause a fight, right?).

Each child gets four pretzel sticks. This problem tells you that

$$12 \div 3 = 4$$

As with multiplication, division also has more than one sign: the *division sign* ($\div$) and the *fraction slash* (/) or *fraction bar* (—). So some other ways to write the same information are

$$^{12}/_3 = 4 \quad \text{and} \quad \frac{12}{3} = 4$$

Whichever way you write it, the idea is the same: When you divide 12 pretzel sticks equally among three people, each person gets four of them.

 When you divide one number by another, the first number is called the *dividend*, the second is called the *divisor*, and the result is the *quotient*. For example, in the division from the earlier example, the dividend is 12, the divisor is 3, and the quotient is 4.

Whatever happened to the division table?

Considering how much time teachers spend on the multiplication table, you may wonder why you've never seen a division table. For one thing, the multiplication table focuses on multiplying all the one-digit numbers by each other. This focus wouldn't work too well for division, because division usually involves at least one number that has more than one digit.

Besides, you can use the multiplication table for division, too, by reversing the way you normally use the table. For example, the multiplication table tells you that $6 \times 7 = 42$. You can reverse this equation to give you these two division problems:

$$42 \div 6 = 7$$
$$42 \div 7 = 6$$

Using the multiplication table in this way takes advantage of the fact that multiplication and division are *inverse operations*. I discuss this important idea further in Chapter 4.

Making short work of long division

In the olden days, knowing how to divide large numbers — for example, 62,997 ÷ 843 — was important. People used *long division,* an organized method for dividing a large number by another number. The process involved dividing, multiplying, subtracting, and dropping numbers down.

But face it — one of the main reasons the pocket calculator was invented was to save 21st-century humans from ever having to do long division again.

Having said that, I need to add that your teacher and math-crazy friends may not agree. Perhaps they just want to make sure you're not completely helpless if your calculator disappears somewhere in your backpack or your desk drawer or the Bermuda Triangle. But if do you get stuck doing page after page of long division against your will, you have my deepest sympathy.

I will go this far, however: Understanding how to do long division with some not-too-horrible numbers is a good idea. In this section, I give you a good start with long division, and with luck, your instructor won't hammer it home with a vengeance!

One-digit divisors

Recall that the *divisor* in a division problem is the number that you're dividing by. When you're doing long division, the size of the divisor is your main concern: Small divisors are easy to work with, and large ones are a royal pain. So I start with a nice, small, one-digit divisor. Suppose you want to find 860 ÷ 5. Start off by writing the problem like this:

$$5 \overline{)860}$$

Unlike the other Big Four operations, long division moves from left to right. In this case, you start with the number in the hundreds column (8). To begin, ask how many times 5 goes into 8 — that is, what's 8 ÷ 5? The answer is 1 (with a little bit left over), so write 1 directly above the 8. Now multiply 1×5 to get 5, place the answer directly below the 8, and draw a line beneath it:

$$5 \overline{)\begin{matrix} 1 \\ 860 \\ 5 \end{matrix}}$$

Subtract 8 – 5 to get 3. (***Note:*** After you subtract, the result should always be smaller than the divisor. If not, you need to write a higher number above the division symbol.) Then bring down the 6 to make the new number 36:

$$5 \overline{)\begin{matrix} 1 \\ 860 \\ -5 \\ \overline{36} \end{matrix}}$$

These steps are one complete cycle, and to complete the problem you just need to repeat them. Now ask how many times 5 goes into 36 — that is, what's 36 ÷ 5? The answer is 7 (with a little left over). Write 7 just above the 6, and then multiply 7×5 to get 35; write the answer under 36:

$$
\begin{array}{r}
17 \\
5\overline{)\,860} \\
-5 \\
\hline
36 \\
35 \\
\hline
\end{array}
$$

Now subtract to get 36 − 35 = 1; bring down the 0 next to the 1 to make the new number 10:

$$
\begin{array}{r}
17 \\
5\overline{)\,860} \\
-5 \\
\hline
36 \\
-35 \\
\hline
10 \\
\end{array}
$$

Another cycle is complete, so begin the next cycle by asking how many times 5 goes into 10 — that is, 10 ÷ 5. The answer this time is 2. Write down the 2 in the answer above the 0. Multiply to get $2 \times 5 = 10$, and write this answer below the 10:

$$
\begin{array}{r}
172 \\
5\overline{)\,860} \\
-5 \\
\hline
36 \\
35 \\
\hline
10 \\
10 \\
\hline
\end{array}
$$

Now subtract 10 − 10 = 0. Because you have no more numbers to bring down, you're finished, and here's the answer (that is, the quotient):

$$
\begin{array}{r}
172 \leftarrow \text{Quotient} \\
5\overline{)\,860} \\
-5 \\
\hline
36 \\
35 \\
\hline
10 \\
-10 \\
\hline
0 \\
\end{array}
$$

So 860 ÷ 5 = 172.

This problem divides evenly, but many don't. The following sections tell you what to do when you run out of numbers to bring down, and Chapter 11 explains how to get a decimal answer.

Getting leftovers: Division with a remainder

Division is different from addition, subtraction, and multiplication in that having a remainder is possible. A *remainder* is simply a portion left over from the division.

The letter *r* indicates that the number that follows is the remainder.

For example, suppose you want to divide seven candy bars between two people without breaking any candy bars into pieces (too messy). So each person receives three candy bars, and one candy bar is left over. This problem shows you the following:

$7 \div 2 = 3$ with a remainder of 1, or 3 r 1

In long division, the remainder is the number that's left when you no longer have numbers to bring down. The following equation shows that $47 \div 3 = 15$ r 2:

$$
\begin{array}{r}
15 \leftarrow \text{Quotient} \\
3\overline{)\ 47} \\
-3 \\
\hline
17 \\
-15 \\
\hline
2 \leftarrow \text{Remainder}
\end{array}
$$

Note that when you're doing division with a small dividend and a large divisor, you always get a quotient of 0 and a remainder of the number you started with:

$1 \div 2 = 0$ r 1

$14 \div 23 = 0$ r 14

$2{,}000 \div 2{,}001 = 0$ r 2,000

Part II
Getting a Handle on Whole Numbers

The 5th Wave   By Rich Tennant

"It's a football/math program. We're tackling multiplication, going long for division, and punting fractions."

In this part . . .

I introduce a whole host of new concepts related to the Big Four operations (adding, subtracting, multiplying, and dividing). I discuss inverse operations, the commutative and associative properties, negative numbers, and inequalities. I tell you all about the three E's of math — expressions, equations, and evaluation. I also show you how to use order of precedence to turn complex-looking math expressions into a single number. You discover how to solve word problems (also called story problems) by writing word equations that make sense of what you're reading. I also give you a bunch of tricks for deciding whether one number is divisible by another and show you how to find the greatest common factor (GCF) and least common multiple (LCM) of two numbers.

Chapter 4

Putting the Big Four Operations to Work

*W*hen you understand the Big Four operations that I cover in Chapter 3 — adding, subtracting, multiplying, and dividing — you can begin to look at math on a whole new level. In this chapter, you extend your understanding of the Big Four operations and move beyond them. I begin by focusing on four important properties of the Big Four operations: inverse operations, commutative operations, associative operations, and distribution. Then I show you how to perform the Big Four on negative numbers.

I continue by introducing you to some important symbols for inequality. Finally, you're ready to move beyond the Big Four by discovering three more-advanced operations: exponents (also called *powers*), square roots (also called *radicals*), and absolute values.

Knowing Properties of the Big Four Operations

When you know how to do the Big Four operations — add, subtract, multiply, and divide — you're ready to grasp a few important *properties* of these important operations. Properties are features of the Big Four operations that always apply no matter what numbers you're working with.

In this section, I introduce you to four important ideas: inverse operations, commutative operations, associative operations, and the distributive property. Understanding these properties can show you hidden connections among the Big Four operations, save you time when calculating, and get you comfortable working with more-abstract concepts in math.

Inverse operations

Each of the Big Four operations has an *inverse* — an operation that undoes it. Addition and subtraction are inverse operations because addition undoes subtraction, and vice versa. For example, here are two inverse equations:

$$1 + 2 = 3$$
$$3 - 2 = 1$$

In the first equation, you start with 1 and add 2 to it, which gives you 3. In the second equation, you have 3 and take away 2 from it, which brings you back to 1. The main idea here is that you're given a starting number — in this case, 1 — and when you add a number and then subtract the same number, you end up again with the starting number. This shows you that subtraction undoes addition.

Similarly, addition undoes subtraction — that is, if you subtract a number and then add the same number, you end up where you started. For example,

$$184 - 10 = 174$$
$$174 + 10 = 184$$

This time, in the first equation, you start with 184 and take away 10 from it, which gives you 174. In the second equation, you have 174 and add 10 to it, which brings you back to 184. In this case, starting with the number 184, when you subtract a number and then add the same number, the addition undoes the subtraction and you end up back at 184.

In the same way, multiplication and division are inverse operations. For example,

$$4 \cdot 5 = 20$$
$$20 \div 5 = 4$$

This time, you start with the number 4 and multiply it by 5 to get 20. And then you divide 20 by 5 to return to where you started at 4. So division undoes multiplication. Similarly,

$$30 \div 10 = 3$$
$$3 \cdot 10 = 30$$

Here, you start with 30, divide by 10, and multiply by 10 to end up back at 30. This shows you that multiplication undoes division.

Commutative operations

Addition and multiplication are both commutative operations. *Commutative* means that you can switch around the order of the numbers without changing the result. This property of addition and multiplication is called the *commutative property*. Here's an example of how addition is commutative:

$3 + 5 = 8$ is the same as $5 + 3 = 8$

If you start out with 5 books and add 3 books, the result is the same as if you start with 3 books and add 5. In each case, you end up with 8 books.

And here's an example of how multiplication is commutative:

$2 \cdot 7 = 14$ is the same as $7 \cdot 2 = 14$

If you have two children and want to give them each seven flowers, you need to buy the same number of flowers as someone who has seven children and wants to give them each two flowers. In both cases, someone buys 14 flowers.

In contrast, subtraction and division are *noncommutative* operations. When you switch around the order of the numbers, the result changes.

Here's an example of how subtraction is noncommutative:

$6 - 4 = 2$ but $4 - 6 = -2$

Subtraction is noncommutative, so if you have $6 and spend $4, the result is *not* the same as if you have $4 and spend $6. In the first case, you still have $2 left over. In the second case, you *owe* $2. That is, switching the numbers around turns the result into a negative number. (I discuss negative numbers later in this chapter.)

And here's an example of how division is noncommutative:

$5 \div 2 = 2 \, r \, 1$ but $2 \div 5 = 0 \, r \, 2$

For example, when you have five dog biscuits to divide between two dogs, each dog gets two biscuits and you have one biscuit left over. But, when you switch the numbers around and try to divide two biscuits among five dogs, you don't have enough biscuits to go around, so each dog gets none and you have two left over.

Associative operations

Addition and multiplication are both *associative operations,* which means that you can group them differently without changing the result. This property of addition and multiplication is also called the *associative property.* Here's an example of how addition is associative. Suppose you want to add 3 + 6 + 2. You can solve this problem in two ways:

$$(3 + 6) + 2 \qquad 3 + (6 + 2)$$

$$= 9 + 2 \qquad\quad = 3 + 8$$

$$= 11 \qquad\qquad = 11$$

In the first case, I start by adding 3 + 6 and then add 2. In the second case, I start by adding 6 + 2 and then add 3. Either way, the sum is 11.

And here's an example of how multiplication is associative. Suppose you want to multiply 5 · 2 · 4. You can solve this problem in two ways:

$$(5 \cdot 2) \cdot 4 \qquad\qquad 5 \cdot (2 \cdot 4)$$

$$= 10 \cdot 4 \qquad\qquad = 5 \cdot 8$$

$$= 40 \qquad\qquad\quad = 40$$

In the first case, I start by multiplying 5 · 2 and then multiply by 4. In the second case, I start by multiplying 2 · 4 and then multiply by 5. Either way, the product is 40. In contrast, subtraction and division are nonassociative operations. This means that grouping them in different ways changes the result.

Don't confuse the commutative property with the associative property. The commutative property tells you that it's okay to *switch around* two numbers that you're adding or multiplying. The associative property tells you that it's okay to *regroup* three numbers using parentheses.

Taken together, the commutative and associative properties allow you to completely rearrange and regroup a string of numbers that you're adding or multiplying without changing the result. You'll find that the freedom to rearrange expressions as you like to be very useful as you move on to algebra in Part V.

Distributing to lighten the load

If you've ever tried to carry a heavy bag of groceries, you may have found that distributing the contents into two smaller bags to be helpful. This same concept also works for multiplication.

In math, *distribution* (also called the *distributive property of multiplication over addition*) allows you to split a large multiplication problem into two smaller ones and add the results to get the answer.

For example, suppose you want to multiply these two numbers:

17×101

You can multiply them out, but distribution provides a different way to think about the problem that you may find easier. Because $101 = 100 + 1$, you can split this problem into two easier problems as follows:

$= 17 \times (100 + 1)$

$= (17 \times 100) + (17 \times 1)$

You take the number outside the parentheses, multiply it by each number inside the parentheses one at a time, then add the products. At this point, you may be able to solve the two multiplications in your head and then add them up easily:

$= 1{,}700 + 17 = 1{,}717$

Distribution becomes even more useful when you get to algebra in Part VI.

Big Four Operations with Negative Numbers

In Chapter 2, I show you how to use the number line to understand how negative numbers work. In this section, I give you a closer look at how to perform the Big Four operations with negative numbers. Negative numbers occur when you subtract a larger number from a smaller one. For example,

$5 - 8 = -3$

In real-world applications, negative numbers are used to represent debt. For example, if you have only five chairs to sell but a customer pays for eight of them, you owe her three more chairs. Even though you may have trouble picturing –3 chairs, you still need to account for this debt, and negative numbers are the right tool for the job.

Addition and subtraction with negative numbers

The great secret to adding and subtracting negative numbers is to turn every problem into a series of ups and downs on the number line. After you know how to do this, you find that all these problems are quite simple.

So in this section, I explain how to add and subtract negative numbers on the number line. Don't worry about memorizing every little bit of this procedure. Instead, just follow along so you get a sense of how negative numbers fit onto the number line. (If you need a quick refresher on how the number line works, see Chapter 1.)

Starting with a negative number

When adding and subtracting on the number line, starting with a negative number isn't much different from starting with a positive number. For example, suppose you want to solve –3 + 4. Using the up and down rules, you get the following:

Start at –3, up 4

So –3 + 4 = 1.

Similarly, suppose you want to solve –2 – 5. Again, the up and down rules help you out. You're subtracting, so move to the left:

Start at –2, down 5

So –2 – 5 = –7.

Adding a negative number

Suppose you want to solve –2 + –4. You already know to start at –2, but where do you go from there? Here's the up and down rule for adding a negative number:

Adding a negative number is the same as subtracting a positive number — that is, go *down* on the number line.

By this rule, –2 + –4 is the same as –2 – 4, so

Start at –2, down 4

So –2 + (–4) = –6.

If you rewrite a subtraction problem as an addition problem — for instance, rewriting 3 – 7 as 3 + (–7) — you can use the commutative and associative properties of addition, which I discuss earlier in this chapter. Just remember to keep the negative sign attached to the number when you rearrange: (–7) + 3.

Subtracting a negative number

The last rule you need to know is how to subtract a negative number. For example, suppose you want to solve 2 – (–3). Here's the up and down rule:

Subtracting a negative number is the same as adding a positive number — that is, *up* on the number line.

This rule tells you that 2 – (–3) is the same as 2 + 3, so

Start at 2, up 3

So 2 – (–3) = 5.

When subtracting negative numbers, you can think of the two minus signs canceling each other out to create a positive.

Multiplication and division with negative numbers

Multiplication and division with negative numbers is virtually the same as it is with positive numbers. The presence of one or more minus signs (–) doesn't change the numerical part of the answer. The only question is whether the sign is positive or negative:

Just remember that when you multiply or divide two numbers,

- ✔ If the numbers have the *same sign,* the result is always positive.
- ✔ If the numbers have *opposite signs,* the result is always negative.

For example,

$$2 \cdot 3 = 6 \qquad 2 \cdot -3 = -6$$
$$-2 \cdot -3 = 6 \qquad -2 \cdot 3 = -6$$

As you can see, the numerical portion of the answer is always 6. The only question is whether the complete answer is 6 or –6. That's where the rule of same or opposite signs comes in.

Another way of thinking of this rule is that the two negatives cancel each other out to make a positive.

Similarly, look at these four division equations:

$$10 \div 2 = 5 \qquad\qquad 10 \div -2 = -5$$
$$-10 \div -2 = 5 \qquad\qquad -10 \div 2 = -5$$

In this case, the numerical portion of the answer is always 5. When the signs are the same, the result is positive, and when the signs are different, the result is negative.

Understanding Units

Anything that can be counted is a *unit.* That's a pretty large category, because almost anything that you can name can be counted. You discover more about units of measurement in Chapter 15. For now, just understand that all units can be counted, which means that you can apply the Big Four operations to units.

Adding and subtracting units

Adding and subtracting units isn't very different from adding and subtracting numbers. Just remember that you can only add or subtract when the units are the same. For example,

3 chairs + 2 chairs = 5 chairs

4 oranges – 1 orange = 3 oranges

What happens when you try to add or subtract different units? Here's an example:

3 chairs + 2 tables = ?

The only way you can complete this addition is to make the units the same:

3 pieces of furniture + 2 pieces of furniture = 5 pieces of furniture

Multiplying and dividing units

You can always multiply and divide units by a *number*. For example, suppose you have four chairs and but find that you need twice as many for a party. Here's how you represent this idea in math:

4 chairs · 2 = 8 chairs

Similarly, suppose you have 20 cherries and want to split them among four people. Here's how you represent this idea:

20 cherries ÷ 4 = 5 cherries

But you have to be careful when multiplying or dividing units by units. For example:

2 apples · 3 apples = ?   WRONG!

12 hats ÷ 6 hats = ?   WRONG!

Neither of these equations makes any sense. In these cases, multiplying or dividing by units is meaningless.

In many cases, however, multiplying and dividing units is okay. For example, multiplying *units of length* (such as inches, miles, or meters) results in *square units*. For example,

3 inches · 3 inches = 9 *square inches*

10 miles · 5 miles = 50 *square miles*

100 meters · 200 meters = 20,000 *square meters*

You find out more about units of length in Chapter 15. Similarly, here are some examples of when dividing units makes sense:

12 slices of pizza ÷ 4 people = 3 slices of pizza/person

140 miles ÷ 2 hours = 70 miles/hour

In these cases, you read the fraction slash (/) as *per*: slices of pizza *per* person or miles *per* hour. You find out more about multiplying and dividing by units in Chapter 15, when I show you how to convert from one unit of measurement to another.

Understanding Inequalities

Sometimes, you want to talk about when two quantities are different. These statements are called *inequalities*. In this section, I discuss four types of inequalities: ≠ (doesn't equal), < (less than), > (greater than), and ≈ (approximately equals).

Doesn't equal (≠)

The simplest inequality is ≠, which you use when two quantities are not equal. For example,

$2 + 2 \neq 5$

$3 \times 4 \neq 34$

$999,999 \neq 1,000,000$

You can read ≠ as "doesn't equal" or "is not equal to." Therefore, read $2 + 2 \neq 5$ as "two plus two doesn't equal five."

Less than (<) and greater than (>)

The symbol < means *less than*. For example, the following statements are true:

4 < 5

100 < 1,000

2 + 2 < 5

Similarly, the symbol > means *greater than*. For example,

5 > 4

100 > 99

2 + 2 > 3

The two symbols < and > are similar and easily confused. Here are two simple ways to remember which is which:

✔ Notice that the < looks sort of like an *L*. This *L* should remind you that it means *less than*.

✔ Remember that in any true statement, the *large* open mouth of the symbol is on the side of the *greater* amount, and the *small* point is on the side of the *lesser* amount.

Approximately equals (≈)

In Chapter 2, I show you how rounding numbers makes large numbers easier to work with. In that chapter, I also introduce ≈, which means *approximately equals*.

For example,

49 ≈ 50

1,024 ≈ 1,000

999,999 ≈ 1,000,000

You can also use ≈ when you estimate the answer to a problem:

1,000,487 + 2,001,932 + 5,000,032

≈ 1,000,000 + 2,000,000 + 5,000,000 = 8,000,000

Beyond the Big Four: Exponents, Square Roots, and Absolute Value

In this section, I introduce you to three new operations that you need as you move on with math: exponents, square roots, and absolute value. As with the Big Four operations, these three operations take numbers and tweak them in various ways.

To tell the truth, these three operations have fewer everyday applications than the Big Four. But you'll be seeing a lot more of them as you progress in your study of math. Fortunately, they aren't difficult, so this is a good time to become familiar with them.

Understanding exponents

Exponents (also called *powers*) are shorthand for repeated multiplication. For example, 2^3 means to multiply 2 by itself 3 times. To do that, use the following notation:

$$2^3 = 2 \cdot 2 \cdot 2 = 8$$

In this example, 2 is the *base number* and 3 is the *exponent.* You can read 2^3 as "2 to the third power" or "2 to the power of 3" (or even "2 cubed," which has to do with the formula for finding the value of a cube — see Chapter 16 for details).

Here's another example:

10^5 means to multiply 10 by itself 5 times

That works out like this:

$$10^5 = 10 \cdot 10 \cdot 10 \cdot 10 \cdot 10 = 100,000$$

This time, 10 is the base number and 5 is the exponent. Read 10^5 as "10 to the fifth power" or "10 to the power of 5."

When the base number is 10, figuring out any exponent is easy. Just write down a 1 and that many 0s after it:

$10^2 = 100$	(1 with two 0s)
$10^7 = 10,000,000$	(1 with seven 0s)
$10^{20} = 100,000,000,000,000,000,000$	(1 with twenty 0s)

Exponents with a base number of 10 are very important in scientific notation, which I cover in Chapter 14.

The most common exponent is the number 2. When you take any whole number to the power of 2, the result is a square number. (For more information on square numbers, see Chapter 1.) For this reason, taking a number to the power of 2 is called *squaring* that number. You can read 3^2 as "three squared," 4^2 as "four squared," and so forth. Here are some squared numbers:

$$3^2 = 3 \cdot 3 = 9$$
$$4^2 = 4 \cdot 4 = 16$$
$$5^2 = 5 \cdot 5 = 25$$

Any number raised to the 0 power equals 1. So 1^0, 37^0, and $999,999^0$ are equivalent, or equal.

Discovering your roots

Earlier in this chapter, in "Knowing Properties of the Big Four Operations," I show you how addition and subtraction are inverse operations. I also show you how multiplication and division are inverse operations. In a similar way, roots are the inverse operation of exponents.

The most common root is the square root. A *square root* undoes an exponent of 2. For example,

$$3^2 = 3 \cdot 3 = 9, \text{ so } \sqrt{9} = 3$$
$$4^2 = 4 \cdot 4 = 16, \text{ so } \sqrt{16} = 4$$
$$5^2 = 5 \cdot 5 = 25, \text{ so } \sqrt{25} = 5$$

You can read the symbol $\sqrt{}$ either as "the square root of" or as "radical." So, read $\sqrt{9}$ as either "the square root of 9" or "radical 9."

As you can see, when you take the square root of any square number, the result is the number that you multiplied by itself to get that square number in the first place. For example, to find $\sqrt{100}$, you ask the question, "What number when multiplied by itself equals 100?" The answer in this case is 10, because

$$10 \cdot 10 = 100, \text{ so } \sqrt{100} = 10$$

You probably won't use square roots too much until you get to algebra, but at that point they become very handy.

Figuring out absolute value

The *absolute value* of a number is the positive value of that number. It tells you how far away from 0 a number is on the number line. The symbol for absolute value is a set of vertical bars.

Taking the absolute value of a positive number doesn't change that number's value. For example,

$|3| = 3$

$|12| = 12$

$|145| = 145$

However, taking the absolute value of a negative number changes it to a positive number:

$|-5| = 5$

$|-10| = 10$

$|-212| = 212$

Chapter 5

A Question of Values: Evaluating Arithmetic Expressions

In this chapter, I introduce you to what I call the Three E's of math: equations, expressions, and evaluation.

The Three E's of Math: Equations, Expressions, and Evaluations

You should find the Three E's of math very familiar because whether you realize it or not, you've been using them for a long time. Whenever you add up the cost of several items at the store, balance your checkbook, or figure out the area of your room, you're evaluating expressions and setting up equations. In this section, I shed light on this stuff and give you a new way to look at it.

You probably already know that an *equation* is a mathematical statement that has an equal sign (=) — for example, 1 + 1 = 2. An *expression* is a string of mathematical symbols that can be placed on one side of an equation — for example, 1 + 1. And *evaluation* is finding out the *value* of an expression as a number — for example, finding out that the expression 1 + 1 is equal to the number 2.

Throughout the rest of the chapter, I show you how to turn expressions into numbers using a set of rules called the *order of operations* (or *order of precedence*). These rules look complicated, but I break them down so you can see for yourself what to do next in any situation.

Equality for All: Equations

An *equation* is a mathematical statement that tells you that two things have the same value — in other words, it's a statement with an equal sign. The equation is one of the most important concepts in mathematics because it allows you to boil down a bunch of complicated information into a single number.

Mathematical equations come in lots of varieties: arithmetic equations, algebraic equations, differential equations, partial differential equations, Diophantine equations, and many more. In this book, I look at only two types: arithmetic equations and algebraic equations.

In this chapter, I discuss only *arithmetic equations,* which are equations involving numbers, the Big Four operations, and the other basic operations I introduce in Chapter 4 (absolute values, exponents, and roots). In Part V, I introduce you to algebraic equations. Here are a few examples of simple arithmetic equations:

$2 + 2 = 4$

$3 \cdot 4 = 12$

$20 \div 2 = 10$

Three properties of equality

Three properties of equality are called *reflexivity*, *symmetry*, and *transitivity*:

- **Reflexivity** says that everything is equal to itself. For example,

  $1 = 1$ $23 = 23$ $1,000,007 = 1,000,007$

- **Symmetry** says that you can switch around the order in which things are equal. For example,

  $4 \cdot 5 = 20$, so $20 = 4 \cdot 5$

- **Transitivity** says that if something is equal to two other things, then those two other things are equal to each other. For example,

  $3 + 1 = 4$ and $4 = 2 \cdot 2$, so $3 + 1 = 2 \cdot 2$

Because equality has all three of these properties, mathematicians call equality an *equivalence relation*. The inequalities that I introduce in Chapter 4 ($\neq$, $>$, $<$, and $\approx$) don't necessarily share all these properties.

And here are a few examples of more-complicated arithmetic equations:

$$1,000 - 1 - 1 - 1 = 997$$
$$(1 \cdot 1) + (2 \cdot 2) = 5$$
$$4^2 - \sqrt{256} = (791 - 842) \cdot 0$$

Hey, it's just an expression

An *expression* is any string of mathematical symbols that can be placed on one side of an equation. Mathematical expressions, just like equations, come in a lot of varieties. In this chapter, I focus only on *arithmetic expressions,* which are expressions that contain numbers, the Big Four operations, and a few other basic operations (see Chapter 4). In Part V, I introduce you to algebraic expressions. Here are a few examples of simple expressions:

$$2 + 2$$
$$-17 + (-1)$$
$$14 \div 7$$

And here are a few examples of more-complicated expressions:

$$(88 - 23) \div 13$$
$$100 + 2 - 3 \cdot 17$$
$$\sqrt{441} + |-2^3|$$

Evaluating the situation

At the root of the word *evaluation* is the word *value.* In other words, when you evaluate something, you find its value. Evaluating an expression is also referred to as *simplifying, solving,* or *finding the value of an expression.* The words may change, but the idea is the same — boiling a string of numbers and math symbols down to a single number.

When you evaluate an arithmetic expression, you simplify it to a single numerical value — that is, you find the number that it's equal to. For example, evaluate the following arithmetic expression:

$$7 \cdot 5$$

How? Simplify it to a single number:

35

Putting the Three E's together

I'm sure you're dying to know how the Three E's — equality, expressions, and evaluation — are all connected. *Evaluation* allows you to take an *expression* containing more than one number and reduce it down to a single number. Then, you can make an *equation,* using an equal sign to connect the expression and the number. For example, here's an *expression* containing four numbers:

1 + 2 + 3 + 4

When you *evaluate* it, you reduce it down to a single number:

10

And now, you can make an *equation* by connecting the expression and the number with an equal sign:

1 + 2 + 3 + 4 = 10

Introducing Order of Operations

When you were a kid, did you ever try putting on your shoes first and then your socks? If you did, you probably discovered this simple rule:

1. **Put on socks.**

2. **Put on shoes.**

Thus you have an order of operations: The socks have to go on your feet before your shoes; so in the act of putting on your shoes and socks, your socks have precedence over your shoes. A simple rule to follow, right?

In this section, I outline a similar set of rules for evaluating expressions called the *order of operations* (sometimes called *order of precedence*). Don't let the long name throw you. Order of operations is just a set of rules to make sure you get your socks and shoes on in the right order, mathematically speaking, so you always get the right answer.

Note: Through most of this book, I introduce overarching themes at the beginning of each section and then explain them later in the chapter instead of building them and finally revealing the result. But order of operations is a bit too confusing to present that way. Instead, I start with a list of four rules and go into more detail about them later in the chapter. Don't let the complexity of these rules scare you off before you work through them!

Evaluate arithmetic expressions from left to right according to the following order of operations:

1. **Parentheses**

2. **Exponents**

3. **Multiplication and division**

4. **Addition and subtraction**

Don't worry about memorizing this list right now. I break it to you slowly in the remaining sections of this chapter, starting from the bottom and working toward the top, as follows:

✔ In "Applying order of operations to Big Four expressions," I show Steps 3 and 4 — how to evaluate expressions with any combination of addition, subtraction, multiplication, and division.

✔ In "Using order of operations in expressions with exponents," I show you how Step 2 fits in — how to evaluate expressions with Big Four operations *plus* exponents, square roots, and absolute value.

✔ In "Understanding order of operations in expressions with parentheses," I show you how Step 1 fits in — how to evaluate all the expressions I explain *plus* expressions with parentheses.

Applying order of operations to Big Four expressions

As I explain earlier in this chapter, evaluating an expression is just simplifying it down to a single number. Now, I get you started on the basics of evaluating expressions that contain any combination of the Big Four operations — adding, subtracting, multiplying, and dividing. (For more on the Big Four, see Chapter 3.) Generally speaking, the Big Four expressions come in the three types outlined in Table 5-1.

Table 5-1	The Three Types of Big Four Expressions	
Expression	*Example*	*Rule*
Contains only addition and subtraction	$12 + 7 - 6 - 3 + 8$	Evaluate left to right.
Contains only multiplication and division	$18 \div 3 \cdot 7 \div 14$	Evaluate left to right.
Mixed-operator expressions: contains a combination of addition/subtraction and multiplication/division	$9 + 6 \div 3$	1. Evaluate multiplication and division left to right. 2. Evaluate addition and subtraction left to right.

In this section, I show you how to identify and evaluate all three types of expressions.

Expressions with only addition and subtraction

Some expressions contain only addition and subtraction. When this is the case, the rule for evaluating the expression is simple.

When an expression contains only addition and subtraction, evaluate it step by step from left to right. For example, suppose you want to evaluate this expression:

$$17 - 5 + 3 - 8$$

Because the only operations are addition and subtraction, you can evaluate from left to right, starting with $17 - 5$:

$$= 12 + 3 - 8$$

As you can see, the number 12 replaces $17 - 5$. Now the expression has three numbers rather than four. Next, evaluate $12 + 3$:

$$= 15 - 8$$

This breaks the expression down to two numbers, which you can evaluate easily:

$$= 7$$

So $17 - 5 + 3 - 8 = 7$.

Expressions with only multiplication and division

Some expressions contain only multiplication and division. When this is the case, the rule for evaluating the expression is pretty straightforward.

When an expression contains only multiplication and division, evaluate it step by step from left to right. Suppose you want to evaluate this expression:

$$9 \cdot 2 \div 6 \div 3 \cdot 2$$

Again, the expression contains only multiplication and division, so you can move from left to right, starting with $9 \cdot 2$:

$= 18 \div 6 \div 3 \cdot 2$

$- 3 \div 3 \cdot 2$

$- 1 \cdot 2$

$= 2$

Notice that the expression shrinks one number at a time until all that's left is 2. So $9 \cdot 2 : 6 \div 3 \cdot 2 = 2$.

Here's another quick example:

$$-2 \cdot 6 \div -4$$

Even though this expression has some negative numbers, the only operations it contains are multiplication and division. So you can evaluate it in two steps from left to right (remembering the rules for multiplying and dividing with negative numbers that I show you in Chapter 4):

$= -12 \div -4$

$= 3$

Thus, $-2 \cdot 6 \div -4 = 3$.

Mixed-operator expressions

Often, an expression contains

- ✔ At least one addition or subtraction operator
- ✔ At least one multiplication or division operator

I call these *mixed-operator expressions*. To evaluate them, you need some stronger medicine. Here's the rule you want to follow.

Evaluate mixed-operator expressions as follows:

1. **Evaluate the multiplication and division from left to right.**
2. **Evaluate the addition and subtraction from left to right.**

For example, suppose you want to evaluate the following expression:

$$5 + 3 \cdot 2 + 8 \div 4$$

As you can see, this expression contains addition, multiplication, and division, so it's a mixed-operator expression. To evaluate it, start out by underlining the multiplication and division in the expression:

$$5 + \underline{3 \cdot 2} + \underline{8 \div 4}$$

Now, evaluate what you underlined from left to right:

$$= 5 + 6 + \underline{8 \div 4}$$
$$= 5 + 6 + 2$$

At this point, you're left with an expression that contains only addition, so you can evaluate it from left to right:

$$= 11 + 2$$
$$= 13$$

Thus, $5 + 3 \cdot 2 + 8 \div 4 = 13$.

Using order of operations in expressions with exponents

Here's what you need to know to evaluate expressions that have exponents (see Chapter 4 for info on exponents).

Evaluate exponents from left to right *before* you begin evaluating Big Four operations (adding, subtracting, multiplying, and dividing).

The trick here is to turn the expression into a Big Four expression and then use what I show you earlier in "Applying order of operations to Big Four expressions." For example, suppose you want to evaluate the following:

$$3 + 5^2 - 6$$

First, evaluate the exponent:

$$= 3 + 25 - 6$$

At this point, the expression contains only addition and subtraction, so you can evaluate it from left to right in two steps:

$$= 28 - 6$$
$$= 22$$

So $3 + 5^2 - 6 = 22$.

Understanding order of precedence in expressions with parentheses

In math, parentheses — ( ) — are often used to group together parts of an expression. When it comes to evaluating expressions, here's what you need to know about parentheses.

To evaluate expressions that contain parentheses,

1. **Evaluate the contents of parentheses, from the inside out.**

2. **Evaluate the rest of the expression.**

Big Four expressions with parentheses

Similarly, suppose you want to evaluate $(1 + 15 \div 5) + (3 - 6) \cdot 5$. This expression contains two sets of parentheses, so evaluate these from left to right. Notice that the first set of parentheses contains a mixed-operator expression, so evaluate this in two steps starting with the division:

$$= (1 + 3) + (3 - 6) \cdot 5$$
$$= 4 + (3 - 6) \cdot 5$$

Now evaluate the contents of the second set of parentheses:

$$= 4 + -3 \cdot 5$$

Now you have a mixed-operator expression, so evaluate the multiplication $(-3 \cdot 5)$ first:

$$= 4 + -15$$

Finally, evaluate the addition:

$$= -11$$

So $(1 + 15 \div 5) + (3 - 6) \cdot 5 = -11$.

Expressions with exponents and parentheses

As another example, try this out:

$$1 + (3 - 6^2 \div 9) \cdot 2^2$$

Start out by working *only* with what's inside the parentheses. The first thing to evaluate there is the exponent, 6^2:

$$= 1 + (3 - 36 \div 9) \cdot 2^2$$

Continue working inside the parentheses by evaluating the division $36 \div 9$:

$$= 1 + (3 - 4) \cdot 2^2$$

Now you can get rid of the parentheses altogether:

$$= 1 + -1 \cdot 2^2$$

At this point, what's left is an expression with an exponent. This expression takes three steps, starting with the exponent:

$$= 1 + -1 \cdot 4$$
$$= 1 + -4$$
$$= -3$$

So $1 + (3 - 6^2 \div 9) \cdot 2^2 = -3$.

Expressions with parentheses raised to an exponent

Sometimes, the entire contents of a set of parentheses are raised to an exponent. In this case, evaluate the contents of the parentheses *before* evaluating the exponent, as usual. Here's an example:

$$(7 - 5)^3$$

First, evaluate $7 - 5$:

$$= 2^3$$

With the parentheses removed, you're ready to evaluate the exponent:

= 8

Once in a rare while, the exponent itself contains parentheses. As always, evaluate what's in the parentheses first. For example,

$21^{(19 + 3 \cdot -6)}$

This time, the smaller expression inside the parentheses is a mixed-operator expression. I underlined the part that you need to evaluate first:

$= 21^{(19 + -18)}$

Now you can finish off what's inside the parentheses:

$= 21^{1}$

At this point, all that's left is a very simple exponent:

= 21

So $21^{(19 + 3 \cdot -6)} = 21$.

Note: Technically, you don't need to put parentheses around the exponent. If you see an expression in the exponent, treat it as though it had parentheses around it. In other words, $21^{19 + 3 \cdot -6}$ means the same thing as $21^{(19 + 3 \cdot -6)}$.

Expressions with nested parentheses

Occasionally, an expression has *nested parentheses:* one or more sets of parentheses inside another set. Here, I give you the rule for handling nested parentheses.

When evaluating an expression with nested parentheses, evaluate what's inside the *innermost* set of parentheses first and work your way toward the *outermost* parentheses.

For example, suppose you want to evaluate the following expression:

$2 + (9 - (\underline{7 - 3}))$

I underlined the contents of the innermost set of parentheses, so evaluate these contents first:

$= 2 + (9 - 4)$

Next, evaluate what's inside the remaining set of parentheses:

$$= 2 + 5$$

Now you can finish things off easily:

$$= 7$$

So $2 + (9 - (7 - 3)) = 7$

As a final example, here's an expression that requires everything from this chapter:

$$4 + (-7 \cdot (2^{(5-1)} - 4 \cdot 6))$$

This expression is about as complicated as you're ever likely to see in pre-algebra: one set of parentheses containing another set, which contains a third set. To start you off, I underlined what's deep inside this third set of parentheses. This is where you begin evaluating:

$$= 4 + (-7 \cdot (2^4 - 4 \cdot 6))$$

Now, what's left is one set of parentheses inside another set. Again, work from the inside out. The smaller expression here is $2^4 - 4 \cdot 6$, so evaluate the exponent first, then the multiplication, and finally the subtraction:

$$= 4 + (-7 \cdot (16 - 4 \cdot 6))$$
$$= 4 + (-7 \cdot (16 - 24))$$
$$= 4 + (-7 \cdot -8)$$

Only one more set of parentheses to go:

$$= 4 + 56$$

At this point, finishing up is easy:

$$= 60$$

Therefore, $4 + (-7 \cdot (2^{(5-1)} - 4 \cdot 6)) = 60$.

As I say earlier in this section, this problem is about as hard as they come at this stage of math. Copy it down and try solving it step by step with the book closed.

Chapter 6

Say What? Turning Words into Numbers

*T*he very mention of word problems — or story problems, as they're sometimes called — is enough to send a cold shiver of terror into the bones of the average math student. Many would rather swim across a moat full of hungry crocodiles than "figure out how many bushels of corn Farmer Brown picked" or "help Aunt Sylvia decide how many cookies to bake." But word problems help you understand the logic behind setting up equations in real-life situations, making math actually useful — even if the scenarios in the word problems you practice on are pretty far-fetched.

In this chapter, I dispel a few myths about word problems. Then, I show you how to solve a word problem in four simple steps. After you understand the basics, I show you how to solve more-complex problems. Some of these problems have longer numbers to calculate, and others may have more complicated stories. In either case, you can see how to work through them step by step.

Dispelling Two Myths about Word Problems

Here are two common myths about word problems:

- ✔ Word problems are always hard.
- ✔ Word problems are only for school — after that, you don't need them.

Both of these ideas are untrue. But they're so common that I want to address them head-on.

Word problems aren't always hard

Word problems don't have to be hard. For example, here's a word problem that you may have run into in first grade:

> Adam had 4 apples. Then Brenda gave him 5 more apples. How many apples does Adam have now?

You can probably do the math in your head, but when you were starting out in math you may have written down:

$$4 + 5 = 9$$

Finally, if you had one of those teachers who made you write out your answer in complete sentences, you wrote "Adam has 9 apples." (Of course, if you were the class clown, you probably wrote, "Adam doesn't have any apples because he ate them all.")

Word problems seem hard when they get too complex to solve in your head and you don't have a system for solving them. In this chapter, I give you a system and show you how to apply it to problems of increasing difficulty. And in Chapters 13, 18, and 24, I give you further practice solving more difficult word problems.

Word problems are useful

In the real world, math rarely comes in the form of equations. It comes in the form of situations that are very similar to word problems.

Whenever you paint a room, prepare a budget, bake a double batch of oatmeal cookies, estimate the cost of a vacation, buy wood to build a shelf, do your taxes, or weigh out the pros and cons of buying a car or leasing one, you need math. And the math skill you need most is understanding how to turn the *situation* you're facing into numbers that you calculate.

Word problems give you practice turning situations — that is, stories — into numbers.

Solving Basic Word Problems

Generally speaking, solving a word problem involves four steps:

1. **Read through the problem and set up *word equations* — that is, equations that contain words as well as numbers.**

2. **Plug in numbers in place of words wherever possible to set up a regular math equation.**

3. **Use math to solve the equation.**

4. **Answer the question that the problem asks.**

Most of this book is about Step 3. This chapter and Chapters 13, 18, and 24 are all about Steps 1 and 2. I show you how to break down a word problem sentence by sentence, jot down the information you need to solve the problem, and then substitute numbers for words to set up an equation.

When you know how to turn a word problem into an equation, the hard part is done. Then you can use the rest of what you find in this book to figure out how to do Step 3 — solve the equation. From there, Step 4 is usually pretty easy, though at the end of each example, I make sure you understand how to do it.

Turning word problems into word equations

The first step to solving a word problem is reading it and putting the information you find into a useful form. In this section, I show you how to squeeze the juice out of a word problem and leave the pits behind!

Jotting down information as word equations

Most word problems give you information about numbers, telling you exactly how much, how many, how fast, how big, and so forth. Here are some examples:

Nunu is spinning 17 plates.

The width of the house is 80 feet.

If the local train is going 25 miles per hour . . .

You need this information to solve the problem. And paper is cheap, so don't be afraid to use it. (If you're concerned about trees, write on the back of all that junk mail you get.) Have a piece of scrap paper handy and jot down a few notes as you read through a word problem.

For example, here's how you can jot down "Nunu is spinning 17 plates":

Nunu = 17

Here's how to note that ". . . the width of the house is 80 feet":

Width = 80

The third example tells you, "If the local train is going 25 miles per hour. . . ." So you can jot down the following:

Local = 25

Don't let the word *if* confuse you. When a problem says "If so-and-so were true . . ." and then asks you a question, assume it *is* true and use this information to answer the question.

When you jot down information this way, you're really turning words into a more useful form called a *word equation*. A word equation has an equal sign like a math equation, but it contains both words and numbers.

Writing relationships: Turning more-complex statements into word equations

When you start doing word problems, you notice that certain words and phrases show up over and over again. For example,

Bobo is spinning five fewer plates than Nunu.

The height of a house is half as long as its width.

The express train is moving three times faster than the local train.

You've probably seen statements such as these in word problems since you were first doing math. Statements like these look like English, but they're really math, so spotting them is important. You can represent each of these types of statements as a word equations that also use Big Four operations. Look again at the first example:

> Bobo is spinning five fewer plates than Nunu.

You don't know the number of plates that either Bobo or Nunu is spinning. But you know that these two numbers are related.

You can express this relationship like this:

> Bobo = Nunu − 5

This word equation is shorter than the statement it came from. And, as you see in the next section, word equations are easy to turn into the math that you need to solve the problem.

Here's another example:

> The height of a house is half as long as its width.

You don't know the width or height of the house, but you know that these numbers are connected.

You can express this relationship between the width and height of the house as the following word equation:

> Height = width ÷ 2

With the same type of thinking, you can express *The express train is moving three times faster than the local train* as this word equation:

> Express = 3 · local

As you can see, each of the examples allows you to set up a word equation using one of the Big Four operations — adding, subtracting, multiplying, and dividing.

Figuring out what the problem's asking

The end of a word problem usually contains the question that you need to answer to solve the problem. You can use word equations to clarify this question so you know right from the start what you're looking for.

For example, you can write the question, "All together, how many plates are Bobo and Nunu spinning?" as

Bobo + Nunu = ?

You can write the question "How tall is the house" as:

Height = ?

Finally, you can rephrase the question "What's the difference in speed between the express train and the local train?" in this way:

Express – Local = ?

Plugging in numbers for words

After you've written out a bunch of word equations, you have the facts you need in a form you can use. Now, you can often solve the problem by plugging numbers from one word equation into another. In this section, I show you how to use the word equations you built in the last section to solve three problems.

Example: Send in the clowns

Some problems involve simple addition or subtraction. Here's an example:

Bobo is spinning five fewer plates than Nunu (Bobo dropped a few). Nunu is spinning 17 plates. All together, how many plates are Bobo and Nunu spinning?

Here's what you have already, just from reading the problem:

Nunu = 17

Bobo + 5 = Nunu

Plugging in the information gives you the following:

Bobo + 5 = ~~Nunu~~ 17

If you see how many plates Bobo is spinning, feel free to jump ahead. If not, here's how you rewrite the addition equation as a subtraction equation (see Chapter 4 for details):

Bobo = 17 – 5 = 12

The problem wants you to find out how many plates the two clowns are spinning together. That is, you need to find out the following:

Bobo + Nunu = ?

Just plug in the numbers, substituting 12 for Bobo and 17 for Nunu:

~~Bobo~~ 12 + ~~Nunu~~ 17 = 29

So Bobo and Nunu are spinning 29 plates.

Example: Our house in the middle of our street

At times, a problem may note relationships that require you to use multiplication or division. Here's an example:

The height of a house is half as long as its width, and the width of the house is 80 feet. How tall is the house?

You already have a head start from what you determined earlier:

Width = 80

Height = Width ÷ 2

You can plug in information as follows, substituting 80 for the word *width*:

Height = ~~Width~~ 80 ÷ 2 = 40

So you know that the height of the house is 40 feet.

Example: I hear the train a comin'

Pay careful attention to what the question is asking. You may have to set up more than one equation. Here's an example:

The express train is moving three times faster than the local train. If the local train is going 25 miles per hour, what's the difference in speed between the express train and the local train?

Here's what you have so far:

Local = 25

Express = 3 · Local

So plug in the information you need:

Express = 3 · 25 ~~Local~~ = 75

In this problem, the question at the end asks you to find the difference in speed between the express train and the local train. Finding the difference between two numbers is subtraction, so here's what you want to find:

Express – Local = ?

You can get what you need to know by plugging in the information you've already found:

~~Express~~ 75 – ~~Local~~ 25 = 50

Therefore, the difference in speed between the express train and the local train is 50 miles per hour.

Solving More-Complex Word Problems

The skills I show you previously in "Solving Basic Word Problems" are important for solving any word problem because they streamline the process and make it simpler. And what's more, you can use those same skills to find your way through more complex problems. Problems become more complex when:

- ✔ The calculations become harder. (For example, instead of a dress costing $30, now it costs $29.95.)
- ✔ The amount of information in the problem increases. (For example, instead of two clowns, now there are five.)

Don't let problems like these scare you. In this section, I show you how to use your new problem-solving skills to solve more difficult word problems.

When numbers get serious

A lot of problems that look tough aren't much more difficult than the problems I show you in the previous sections. For example, consider this problem:

Aunt Effie has $732.84 hidden in her pillowcase, and Aunt Jezebel has $234.19 less than Aunt Effie has. How much money do the two women have all together?

One question you may have is how these women ever get any sleep with all that change clinking around under their heads. But moving on to the math, even though the numbers are larger, the principle is still the same as in problems in the earlier sections. Start reading from the beginning: "Aunt Effie has \$732.84. . . ." This text is just information to jot down as a simple word equation:

Effie = \$732.84

Continuing, you read: ". . . Aunt Jezebel has *\$234.19 less than* Aunt Effie has." It's another statement you can write as a word equation:

Jezebel = Effie − \$234.19

Now you can plug in the number \$732.84 where you see Aunt Effie's name in the equation:

Jezebel − ~~Effie~~ \$732.84 − \$234.19

So far, the big numbers haven't been any trouble. At this point, though, you probably need to stop to do the subtraction:

$$\begin{array}{r} \$732.84 \\ -\ \underline{\$234.19} \\ \$498.65 \end{array}$$

Now you can jot this information down as always:

Jezebel = \$498.65

The question at the end of the problem asks you to find out how much money the two women have all together. Here's how to represent this question as an equation:

Effie + Jezebel = ?

You can plug information into this equation:

~~Effie~~ \$732.84 + ~~Jezebel~~ \$498.65 = ?

Again, because the numbers are large, you probably have to stop to do the math:

$$\begin{array}{r} \$732.84 \\ +\ \underline{\$498.65} \\ \$1,231.49 \end{array}$$

So, all together, Aunt Effie and Aunt Jezebel have $1,231.49.

As you can see, the procedure for solving this problem is basically the same as for the simpler problems in the earlier sections. The only difference is that you have to stop to do some addition and subtraction.

Too much information

When the going gets tough, knowing the system for writing word equations really becomes helpful. Here's a word problem that's designed to scare you off — but with your new skills, you're ready for it:

> Four women collected money to save the endangered Salt Creek tiger beetle. Keisha collected $160, Brie collected $50 more than Keisha, Amy collected twice as much as Brie, and together Amy and Sophia collected $700. How much money did the four women collect all together?

If you try to do this problem all in your head, you'll probably get confused. Instead, take it line by line and just jot down word equations as I discuss earlier in this chapter.

First, "Keisha collected $160." So jot down the following:

Keisha = 160

Next, "Brie collected $50 dollars more than Keisha," so write

Brie = Keisha + 50

After that, "Amy collected twice as much as Brie":

Amy = Brie · 2

And finally, "together, Amy and Sophia collected $700":

Amy + Sophia = 700

That's all the information that the problem gives you, so now you can start working with it. Keisha collected $160, so you can plug in 160 anywhere you find Keisha's name:

Brie = ~~Keisha~~ 160 + 50 = 210

Now you know how much Brie collected, so you can plug this information into the next equation:

Amy = ~~Brie~~ 210 · 2 = 420

This equation tells you how much Amy collected, so you can plug this number into the last equation:

~~Amy~~ 420 + Sophia = 700

To solve this problem, change it from addition to subtraction using inverse operations, as I show you in Chapter 4:

Sophia = 700 − 420 = 280

Now that you know how much money each woman collected, you can answer the question at the end of the problem:

Keisha + Brie + Amy + Sophia = ?

You can plug in this information easily:

~~Keisha~~ 160 + ~~Brie~~ 210 + ~~Amy~~ 420 + ~~Sophia~~ 280 = 1,070

So you can conclude that the four women collected $1,070 all together.

Putting it all together

Here's one final example putting together everything from this chapter. Try writing down this problem and working it through step by step on your own. If you get stuck, come back here. When you can solve it from beginning to end with the book closed, you'll have a good grasp of how to solve word problems:

On a recent shopping trip, Travis bought six shirts for $19.95 each and two pairs of pants for $34.60 each. He then bought a jacket that cost $37.08 less than he paid for both pairs of pants. If he paid the cashier with three $100 bills, how much change did he receive?

On the first read-through, you may wonder how Travis found a store that prices jackets that way. Believe me — it was quite a challenge. Anyway, back to the problem. You can jot down the following word equations:

Shirts = $19.95 · 6

Pants = $34.60 · 2

Jacket = Pants − $37.08

The numbers in this problem are probably longer than you can solve in your head, so they require some attention:

$$\begin{array}{r} \$19.95 \\ \times\ \ \ \ 6 \\ \hline \$119.70 \end{array} \qquad \begin{array}{r} \$34.60 \\ \times\ \ \ \ 2 \\ \hline \$69.20 \end{array}$$

With this done, you can fill in some more information:

Shirts = $119.70

Pants = $69.20

Jacket = Pants − $37.08

Now you can plug in $69.20 for *pants:*

Jacket = ~~Pants~~ $69.20 − $37.08

Again, because the numbers are long, you need to solve this equation separately:

$$\begin{array}{r} \$69.20 \\ -\ \$37.08 \\ \hline \$32.12 \end{array}$$

This equation gives you the price of the jacket:

Jacket = $32.12

Now that you have the price of the shirts, pants, and jacket, you can find out how much Travis spent:

Amount Travis spent = ~~Shirts~~ $119.70 + ~~Pants~~ $69.20 + ~~Jacket~~ $32.12

Again, you have another equation to solve:

$$\begin{array}{r} \$119.70 \\ \$\ \ 69.20 \\ +\ \$\ \ 32.12 \\ \hline \$221.02 \end{array}$$

So you can jot down the following:

Amount Travis spent = $221.02

The problem is asking you to find out how much change Travis received from $300, so jot this down:

Change = $300 − Amount Travis spent

You can plug in the amount that Travis spent:

Change = $300 − $221.02

And do just one more equation:

$300.00
− $221.02
$ 78.98

So you can jot down the answer:

Change = $78.98

Therefore, Travis received $78.98 in change.

Chapter 7

Divisibility

In This Chapter

▶ Finding out whether a number is divisible by 2, 3, 5, 9, 10, or 11

▶ Seeing the difference between prime numbers and composite numbers

When one number is *divisible* by another, you can divide the first number by the second number without getting a remainder (see Chapter 3 for details on division). In this chapter, I explore divisibility from a variety of angles.

To start out, I show you a bunch of handy tricks for discovering whether one number is divisible by another without actually doing the division. (In fact, there's no long division anywhere in this chapter!) After that, I talk about prime numbers and composite numbers (which I introduce briefly in Chapter 1).

This discussion, plus what follows in Chapter 8, can help make your encounter with fractions in Part III a lot friendlier.

Knowing the Divisibility Tricks

As you begin to work with fractions in Part III, the question of whether one number is divisible by another comes up a lot. In this section, I give you a bunch of time-saving tricks for finding out whether one number is divisible by another without actually making you do the division.

Counting everyone in: Numbers you can divide everything by

Every number is divisible by 1. As you can see, when you divide any number by 1, the answer is the number itself, with no remainder:

$$2 \div 1 = 2$$
$$17 \div 1 = 17$$
$$431 \div 1 = 431$$

Similarly, every number (except 0) is divisible by itself. Clearly, when you divide any number by itself, the answer is 1:

$$5 \div 5 = 1$$
$$28 \div 28 = 1$$
$$873 \div 873 = 1$$

You can't divide any number by 0. Mathematicians say that dividing by 0 is *undefined*.

In the end: Looking at the final digits

You can tell whether a number is divisible by 2, 5, 10, 100, or 1,000 simply by looking at how the number ends. No calculations required.

Divisible by 2

Every even number — that is, every number that ends in 2, 4, 6, 8, or 0 — is divisible by 2. For example, the following bolded numbers are divisible by 2:

6 ÷ 2 = 3	**538** ÷ 2 = 269	**77,144** ÷ 2 = 38,572
22 ÷ 2 = 11	**6,790** ÷ 2 = 3,395	**212,116** ÷ 2 = 106,058

Divisible by 5

Every number that ends in either 5 or 0 is divisible by 5. The following bolded numbers are divisible by 5:

15 ÷ 5 = 3	**6,970** ÷ 5 = 1,394	**511,725** ÷ 5 = 102,345
625 ÷ 5 = 125	**44,440** ÷ 5 = 8,888	**9,876,630** ÷ 5 = 1,975,326

Divisible by 10, 100, or 1,000

Every number that ends in 0 is divisible by 10. The following bolded numbers are divisible by 10:

20 ÷ 10 = 2	**170** ÷ 10 = 17	**56,720** ÷ 10 = 5,672

Every number that ends in 00 is divisible by 100:

300 $\div$ 100 = 3   **8,300** $\div$ 100 = 83   **634,900** $\div$ 100 = 6,349

And every number that ends in 000 is divisible by 1,000:

6,000 $\div$ 1,000 = 6   **99,000** $\div$ 1,000 = 99   **1,234,000** $\div$ 1,000 = 1,234

In general, every number that ends with a string of 0s is divisible by the number you get when you write 1 followed by that many 0s. For example,

900,000 is divisible by 100,000

235,000,000 is divisible by 1,000,000

820,000,000,000 is divisible by 10,000,000,000

When numbers start to get this large, mathematicians usually switch over to *scientific notation* to write them more efficiently. In Chapter 14, I show you all about how to work with scientific notation.

Add it up: Checking divisibility by adding up digits

Sometimes you can check divisibility by adding up all or some of the digits in a number. The sum of a number's digits is called its *digital roots*. Finding the digital root of a number is easy, and it's a handy thing to know.

To find the digital root of a number, just add up the digits and repeat this process until you get a one-digit number. Here are some examples:

The digital root of 24 is 6 because 2 + 4 = 6.

The digital root of 143 is 8 because 1 + 4 + 3 = 8.

The digital root of 51,111 is 9 because 5 + 1 + 1 + 1 + 1 = 9.

Sometimes, you need to do this process more than once. Here's how to find the digital root of the number 87,482. You have to repeat the process three times, but eventually you find that the digital root of 87,482 is 2:

8 + 7 + 4 + 8 + 2 = 29

2 + 9 = 11

1 + 1 = 2

Read on to find out how sums of digits can help you check for divisibility by 3, 9, or 11.

Divisible by 3

Every number whose digital root is 3, 6, or 9 is divisible by 3.

First find the digital root of a number by adding its digits until you get a single-digit number. Here are the digital roots of 18, 51, and 975:

18:     $1 + 8 = 9$

51:     $5 + 1 = 6$

975:    $9 + 7 + 5 = 21; 2 + 1 = 3$

With the numbers 18 and 51, adding the digits leads immediately to digital roots 9 and 6, respectively. With 975, when you add up the digits, you first get 21, so you then add up the digits in 21 to get the digital root 3. Thus, these three numbers are all divisible by 3. If you do the actual division, you find that $18 \div 3 = 6$, $51 \div 3 = 17$, and $975 \div 3 = 325$, so the method checks out.

However, when the digital root of a number is anything other than 3, 6, or 9, the number *isn't* divisible by 3:

1,037: $1 + 0 + 3 + 7 = 11; 1 + 1 = 2$

Because the digital root of 1,037 is 2, 1,037 *isn't* divisible by 3. If you try to divide by 3, you end up with 345 r 2.

Divisible by 9

Every number whose digital root is 9 is divisible by 9.

To test whether a number is divisible by 9, find its digital root by adding up its digits until you get a one-digit number. Here are some examples:

36:     $3 + 6 = 9$

243:    $2 + 4 + 3 = 9$

7,587: $7 + 5 + 8 + 7 = 27; 2 + 7 = 9$

With the numbers 36 and 243, adding the digits leads immediately to digital roots of 9 in both cases. With 7,587, however, when you add up the digits, you get 27, so you then add up the digits in 27 to get the digital root 9. Thus, all three of these numbers are divisible by 9. You can verify this by doing the division: $36 \div 9 = 4$, $243 \div 9 = 27$, and $7,857 \div 9 = 873$.

However, when the digital root of a number is anything other than 3, 6, or 9, the number isn't divisible by 3. Here's an example:

706:   7 + 0 + 6 = 13; 1 + 3 = 4

Because the digital root of 706 is 4, 706 *isn't* divisible by 9. If you try to divide 706 by 9, you get 78 r 4.

Divisible by 11

Two-digit numbers that are divisible by 11 are hard to miss because they simply repeat the same digit twice. Here are all the numbers less than 100 that are divisible by 11:

11   22   33   44   55   66   77   88   99

For numbers between 100 and 200, use this rule: Every three-digit number whose first and third digits add up to its second digit is divisible by 11. For example, suppose you want to decide whether the number 154 is divisible by 11. Just add up the first and third digit:

1 + 4 = 5

Because these two numbers add up to the second digit, 5, the number 154 is divisible by 11. If you divide, you get 154 ÷ 11 = 14, a whole number.

Now suppose you want to figure out whether 136 is divisible by 11. Add the first and third digits:

1 + 6 = 7

Because the first and third digits add up to 7 instead of 3, the number 136 isn't divisible by 11. You can find that 136 ÷ 11 = 12 r 4.

For numbers of any length, the rule is slightly more complicated, but it's still often easier than doing long division. A number is divisible by 11 when its *alternate* digits

✔ Add up to the same number *or*

✔ Add up to two numbers that, when one is subtracted from the other, result in a number that's divisible by 11

For example, suppose you want to discover whether the number 15,983 is divisible by 11. To start out, underline alternate digits (every other digit):

15,983

Now add up the underlined digits and the non-underlined digits:

$$1 + \underline{9} + \underline{3} = \underline{13}$$

$$15,\underline{9}8\underline{3}$$

$$5 + 8 = 13$$

Because these two sets of digits both add up to 13, the number 15,983 is divisible by 11. If you check the division, $15,983 \div 11 = 1,453$.

Now suppose you want to find out whether 9,181,909 is divisible by 11. Again, underline alternate digits and add up the two groups:

$$\underline{9} + \underline{8} + \underline{9} + \underline{9} = \underline{35}$$

$$\underline{9},1\underline{8}1,\underline{9}0\underline{9}$$

$$1 + 1 + 0 = 2$$

Obviously, 35 and 2 aren't equal. But notice that $35 - 2 = 33$. Because 33 is divisible by 11, the number 9,181,909 is also divisible by 11. The actual answer is $9,181,909 \div 11 = 834,719$

Identifying Prime and Composite Numbers

In the earlier section titled "Counting everyone in: Numbers you can divide everything by," I show you that every number (except 0 and 1) is divisible by at least two numbers: 1 and itself. In this section, I explore prime numbers and composite numbers (which I introduce you to in Chapter 1).

In Chapter 8, you need to know how to tell prime numbers from composite in order to break a number down into its prime factors. This, in turn, is important when you begin working with fractions.

A *prime number* is divisible by exactly two positive whole numbers: 1 and the number itself. A *composite number* is divisible by at least three numbers.

For example, 2 is a prime number because when you divide it by any number but 1 and 2, you get a remainder. So there's only one way to multiply two counting numbers together and get 2 as a product:

$$1 \cdot 2 = 2$$

Similarly, 3 is prime because when you divide by any number but 1 or 3, you get a remainder. So the only way to multiply two numbers together and get 3 as a product is the following:

$1 \cdot 3 = 3$

On the other hand, 4 is a composite number because it's divisible by three numbers: 1, 2, and 4. In this case, you have two ways to multiply two counting numbers and get a product of 4:

$1 \cdot 4 = 4$

$2 \cdot 2 = 4$

But 5 is a prime number, because it's divisible only by 1 and 5. Here's the only way to multiply two counting numbers together and get 5 as a product:

$1 \cdot 5 = 5$

And 6 is a composite number because it's divisible by 1, 2, 3, and 6. Here are two ways to multiply two counting numbers and get a product of 6:

$1 \cdot 6 = 6$

$2 \cdot 3 = 6$

Every counting number except 1 is either prime or composite. The reason 1 is neither is that it's divisible by only *one* number, which is 1.

Here's a list of the prime numbers that are less than 30:

2, 3, 5, 7, 11, 13, 17, 19, 23, 29

Remember the first four prime numbers: 2, 3, 5, and 7. Every composite number less than 100 is divisible by at least one of these numbers. This fact makes it easy to test whether a number under 100 is prime: Simply test it for divisibility by 2, 3, 5, and 7. If it's divisible by any of these numbers, it's composite — if not, it's prime.

For example, suppose you want to find out whether the number 79 is prime or composite without actually doing the division. Here's how you think it out, using the tricks I show you earlier in "Knowing the Divisibility Tricks":

- ✔ 79 is an odd number, so it isn't divisible by 2.
- ✔ 79 has a digital root of 8 (because 7 + 9 = 16; 1 + 6 = 7), so it isn't divisible by 3.

✔ 79 doesn't end in 5 or 0, so it isn't divisible by 5.

✔ Even though there's no trick for divisibility by 7, you know that 77 is divisible by 7. So $79 \div 7$ would leave remainder of 2, which tells you that 79 isn't divisible by 7.

Because 79 is less than 100 and isn't divisible by 2, 3, 5, or 7, you know that 79 is a prime number.

Now test whether 93 is prime or composite:

✔ 93 is an odd number, so it isn't divisible by 2.

✔ 93 has a digital root of 3 (because $9 + 3 = 12$ and $1 + 2 = 3$), so 93 is divisible by 3.

You don't need to look further. Because 93 is divisible by 3, you know it's composite.

Chapter 8

Fabulous Factors and Marvelous Multiples

....

In This Chapter

▶ Understanding how factors and multiples are related

▶ Listing all the factors of a number

▶ Breaking a number down into its prime factors

▶ Generating multiples of a number

▶ Finding the greatest common factor (GCF) and least common multiple (LCM)

....

In Chapter 2, I introduce you to sequences of numbers based on the multiplication table. In this chapter, I tell you about two important ways to think about these sequences: as *factors* and as *multiples*. Factors and multiples are really two sides of the same coin. Here I show you what you need to know about these two important concepts.

For starters, I show you how to decompose (split up) any number into its prime factors. Along the way, I give you a bunch of handy tricks for finding out whether a number is a factor of another number. To finish up on factors, I show you how to find the greatest common factor (GCF) of any set of numbers. After that, I tackle multiples, showing you a couple of ways to find the least common multiple of a set of numbers.

Knowing Six Ways to Say the Same Thing

In this section, I introduce you to factors and multiples, and I show you how these two important concepts are connected. As I discuss in Chapter 4, multiplication and division are inverse operations. For example, the following equation is true:

$$5 \cdot 4 = 20$$

So this inverse equation is also true:

$$20 \div 4 = 5$$

You may have noticed that in math, you tend to run into the same ideas over and over again. For example, mathematicians have six different ways to talk about this relationship.

The following three statements all focus on the relationship between 5 and 20 from the perspective of multiplication:

- 5 *multiplied* by some number is 20
- 5 is a *factor* of 20
- 20 is a *multiple* of 5

In two of the examples, you can see this relationship reflected in the words *multiplied* and *multiple*. For the remaining example, keep in mind that two *factors* are multiplied to equal a product.

Similarly, the following three statements all focus on the relationship between 5 and 20 from the perspective of division:

- 20 *divided* by some number is 5
- 20 is *divisible by* 5
- 5 is a *divisor* of 20

Why do mathematicians need all these words for the same thing? Maybe for the same reason that Eskimos need a bunch of words for snow. In any case, in this chapter, I focus on the words *factor* and *multiple*. When you understand the concepts, which word you choose doesn't matter a whole lot.

Connecting Factors and Multiples

When one number is a factor of a second number, the second number is a multiple of the first number. For example, 20 is divisible by 5, so

- 5 is a factor of 20
- 20 is a multiple of 5

Don't mix which number is the factor and which is the multiple. The factor is always the lower number and the multiple is always the higher number.

If you have trouble remembering which number is the factor and which is the multiple, jot them down in order from lowest to highest and write the letters F and M in alphabetical order under them.

For example, 10 divides 40 evenly, so jot down:

    10   40
    F   M

This setup should remind you that 10 is a factor of 40 and that 40 is a multiple of 10.

Fabulous Factors

In this section, I introduce you to factors. First, I show you how to find out whether one number is a factor of another. Then I show you how to list a number's factors. After that, I introduce the key idea of a number's prime factors. This information all leads up to an essential skill: finding the greatest common factor (GCF) of a set of numbers.

Deciding when one number is a factor of another

You can easily tell whether a number is a factor of a second number: Just divide the second number by the first. If it divides evenly (with no remainder), the number is a factor; otherwise, it's not a factor.

For example, suppose you want to know whether 7 is a factor of 56. Here's how you find out:

$$56 \div 7 = 8$$

Because 7 divides 56 without leaving a remainder, 7 is a factor of 56.

And here's how you find out whether 4 is a factor of 34:

$$34 \div 4 = 8 \text{ r } 2$$

Because 4 divides 34 with a remainder of 2, 4 isn't a factor of 34.

This method works no matter how large the numbers are.

Some teachers use factoring problems to test you on long division. For a refresher on how to do long division, see Chapter 3.

Generating a number's factors

The greatest factor of any number is the number itself, so you can always list all the factors of any number because you have a stopping point. Here's how to list all the factors of a number:

1. **Begin the list with 1, leave some space for other numbers, and end the list with the number itself.**

2. **Test whether 2 is a factor — that is, see whether the number is divisible by 2 (for tricks on testing for divisibility, see Chapter 7).**

   If it is, add 2 to the list, along with the original number divided by 2 as the second-to-last number on the list.

3. **Test the number 3 in the same way.**

4. **Continue testing numbers until the beginning of the list meets the end of the list.**

An example should help make this clear. Suppose you want to list all the factors of the number 18. According to Step 1, begin the list with 1 and end it with 18:

> 1   ...   18

Remember from Chapter 7 that every number — whether prime or composite — is divisible by itself and 1. So automatically, 1 and 18 are both factors of 18.

Next, test whether the number 2 is a factor of 18:

> $18 \div 2 = 9$

Because 2 divides 18 without a remainder, 2 is a factor of 18. (For a bunch of easy divisibility tricks, check out Chapter 3.) So both 2 *and* 9 are factors of 18, and you can add them both to the list:

> 1   2   ...   9   18

Notice that I add 9 as the second-to-last number of the list. Doing this reminds you that you don't have to check any number higher than 9.

Now test 3 in the same way:

$$18 \div 3 = 6$$

So both 3 *and* 6 are factors of 18:

1	2	3	...	6	9	18

At this point, you're almost done. You have to check only the numbers between 3 and 6 — that is, the numbers 4 and 5:

$$18 \div 4 = 4 \text{ r } 2$$
$$18 \div 5 = 3 \text{ r } 3$$

Thus, neither 4 nor 5 is a factor of 18, so this is a complete list of the factors of 18:

1	2	3	6	9	18

Prime factors

In Chapter 7, I discuss prime numbers and composite numbers. A prime number is divisible only by 1 and itself — for example, the number 7 is divisible only by 1 and 7. On the other hand, a composite number is divisible by at least one number other than 1 and itself — for example, the number 9 is divisible not only by 1 and 9 but also by 3.

A number's *prime factors* are the set of prime numbers (including repeats) that equal that number when multiplied together. For example, here are the prime factors of the numbers 10, 30, and 72:

$$10 = 2 \cdot 5$$
$$30 = 2 \cdot 3 \cdot 5$$
$$72 = 2 \cdot 2 \cdot 2 \cdot 3 \cdot 3$$

In the last example, the prime factors of 72 include the number 2 repeated three times and the number 3 repeated twice.

REMEMBER

The best way to break a composite number down into its prime factors is using a *factorization tree*. Here's how it works:

1. **Split the number into any two factors and check off the original number.**

2. **If either of these factors is prime, circle it.**

3. **Repeat steps 1 and 2 for any number that is neither circled nor checked.**

4. **When every number in the tree is either checked or circled, the tree is finished, and the circled numbers are the prime factors of the original number.**

For example, to break down the number 56 into its prime factors, start by finding two numbers (other than 1 or 56) that, when multiplied, give you a product of 56. In this case, remember that 7 · 8 = 56. See Figure 8-1.

Figure 8-1:
Finding two factors of 56; 7 is prime.

As you can see, I break 56 down into two factors and check it off. I also circle 7 because it's a prime number. Now 8 is a neither checked nor circled, so I repeat the process, as shown in Figure 8-2.

Figure 8-2:
Continuing the number breakdown with 8.

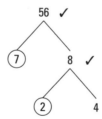

This time, I break 8 into two factors (2 · 4 = 8) and check it off. This time, 2 is prime, so I circle it. But 4 remains unchecked and uncircled, so I continue with Figure 8-3.

At this point, every number in the tree is either circled or checked, so the tree is finished. The four circled numbers — 2, 2, 2, and 7 — are the prime factors of 56. To check this result, just multiply the prime factors together:

$$2 \cdot 2 \cdot 2 \cdot 7 = 56$$

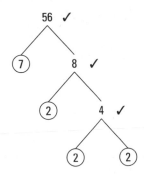

You can see why this is called a tree: Starting at the top, the numbers tend to branch off like an upside-down tree.

What happens when you try to build a tree starting with a prime number — for example, 7? Well, you don't have to go very far (see Figure 8-4).

That's it — you're done! This example shows you that every prime number is its own prime factor.

Here's a list of numbers less than 20 with their prime factorizations. (As you find out in Chapter 2, 1 is neither prime nor composite, so it doesn't have a prime factorization.)

2	$8 = 2 \cdot 2 \cdot 2$	$14 = 2 \cdot 7$
3	$9 = 3 \cdot 3$	$15 = 3 \cdot 5$
$4 = 2 \cdot 2$	$10 = 2 \cdot 5$	$16 = 2 \cdot 2 \cdot 2 \cdot 2$
5	11	17
$6 = 2 \cdot 3$	$12 = 2 \cdot 2 \cdot 3$	$18 = 2 \cdot 3 \cdot 3$
7	13	19

As you can see, the eight prime numbers that I list here are their own prime factorizations. The remaining numbers are composite, so they can all be broken down into smaller prime factors.

Every number has a unique prime factorization. This fact is important — so important that it's called the Fundamental Theorem of Arithmetic. In a way, a number's prime factorization is like its fingerprint — a unique and foolproof way to identify a number.

Knowing how to break a number down to its prime factorization is a handy skill to have. Using the factorization tree allows you to *factor out* one number after another until all you're left with are primes.

Finding prime factorizations for numbers 100 or less

When you build a factorization tree, the first step is usually the hardest. That's because as you proceed, the numbers get smaller and easier to work with. With fairly small numbers, the factorization tree is usually easy to use.

As the number you're trying to factor increases, you may find the first step to be a little more difficult. This is especially true when you don't recognize the number from the multiplication table. The trick is to find someplace to start.

Whenever possible, factor out 5s and 2s first. As I discuss in Chapter 7, you can easily tell when a number is divisible by 2 or by 5.

For example, suppose you want the prime factorization of the number 84. Because you know that 84 is divisible by 2, you can factor out a 2, as shown in Figure 8-5.

Figure 8-5:
Factoring out 2 from 84.

At this point, you should recognize 42 from the multiplication table $(6 \cdot 7 = 42)$.

This tree is now easy to complete (see Figure 8-6).

Figure 8-6:
Completing the factoring of 84.

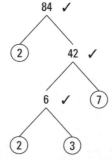

The resulting prime factorization for 84 is as follows:

$$84 = 2 \cdot 7 \cdot 2 \cdot 3$$

If you like, though, you can rearrange the factors from lowest to highest:

$$84 = 2 \cdot 2 \cdot 3 \cdot 7$$

By far, the most difficult situation occurs when you're trying to find the prime factors of a prime number but don't know it. For example, suppose you want to find the prime factorization for the number 71. This time, you don't recognize the number from the multiplication tables and it isn't divisible by 2 or 5. What next?

If a number that's less than 100 (actually, less than 121) isn't divisible by 2, 3, 5, or 7, it's a prime number.

Testing for divisibility by 3 by finding the digital root of 71 (that is, by adding the digits) is easy. As I explain in Chapter 7, numbers divisible by 3 have digital roots of 3, 6, or 9.

$$7 + 1 = 8$$

Because the digital root of 71 is 8, 71 isn't divisible by 3. Divide to test whether 71 is divisible by 7:

$$71 \div 7 = 10 \text{ r } 1$$

So now you know that 71 isn't divisible by 2, 3, 5, or 7. Therefore, 71 is a prime number, so you're done.

Finding prime factorizations for numbers greater than 100

Most of the time, you don't have to worry about finding the prime factorizations of numbers greater than 100. Just in case, though, here's what you need to know.

As I mention in the preceding section, factor out the 5s and 2s first. A special case is when the number you're factoring ends in one or more 0s. In this case, you can factor out a 10 for every 0. For example, Figure 8-7 shows the first step.

Figure 8-7:
The first step in factoring 700.

After you do the first step, the rest of the tree becomes very easy (see Figure 8-8):

Figure 8-8:
Completing
the factoring
of 700.

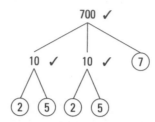

This shows that the prime factorization of 700 is

$$700 = 2 \cdot 2 \cdot 5 \cdot 5 \cdot 7$$

If the number isn't divisible by either 2 or 5, use your divisibility trick for 3 (see Chapter 7) and factor out as many 3s as you can. Then factor out 7s if possible (sorry, I don't have a trick for 7s) and finally 11s.

If a number that's less than 289 isn't divisible by 2, 3, 5, 7, 11, or 13, it's prime. As always, every prime number is its own prime factorization, so when you know that a number is prime, you're done. Most of the time, with larger numbers, a combination of tricks can handle the job.

Finding the greatest common factor (GCF)

After you understand how to find the factors of a number (see "Generating a number's factors"), you're ready to move on to the main event: finding the greatest common factor of several numbers.

The *greatest common factor* (GCF) of a set of numbers is the largest number that's a factor of all those numbers. For example, the GCF of the numbers 4 and 6 is 2 because 2 is the greatest number that's a factor of both 4 and 6. In this section, I show you two ways to find the GCF.

Using a list of factors to find the GCF

The first method for finding the GCF is quicker when you're dealing with smaller numbers.

To find the GCF of a set of numbers, list all the factors of each number, as I show you in "Generating a number's factors." The greatest factor appearing on every list is the GCF. For example, to find the GCF of 6 and 15, first list all the factors of each number.

Factors of 6: 1, 2, 3, 6

Factors of 15: 1, 3, 5, 15

Because 3 is the greatest factor that appears on both lists, 3 is the GCF of 6 and 15.

As another example, suppose you want to find the GCF of 9, 20, and 25. Start by listing the factors of each:

Factors of 9: 1, 3, 9

Factors of 20: 1, 2, 4, 5, 10, 20

Factors of 25: 1, 5, 25

In this case, the only factor that appears on all three lists is 1, so 1 is the GCF of 9, 20, and 25.

Using prime factorization to find the GCF

You can use prime factorization to find the GCF of a set of numbers. This often works better for large numbers, where generating lists of all factors can be time-consuming.

Here's how to find the GCF of a set of numbers using prime factorization:

1. **List the prime factors of each number (see the earlier "Prime factors" section).**

2. **Circle every common prime factor — that is, every prime factor that's a factor of every number in the set.**

3. **Multiply all the circled numbers.**

   The result is the GCF.

For example, suppose you want to find the GCF of 28, 42, and 70. Step 1 says to list the prime factors of each number. Step 2 says to circle every prime factor that's common to all three numbers (as shown in Figure 8-9).

Figure 8-9:
Finding the
GCF of 28,
42, and 70.

$28 = ②\ 2\ ⑦$

$42 = ②\ 3\ ⑦$

$70 = ②\ 5\ ⑦$

As you can see, the numbers 2 and 7 are common factors of all three numbers. Multiply these circled numbers together:

$2 \cdot 7 = 14$

Thus, the GCF of 28, 42, and 70 is 14.

Knowing how to find the GCF of a set of numbers is important when you begin reducing fractions to lowest terms. (For more on reducing fractions, see Chapter 9.)

Marvelous Multiples

Even though multiples tend to be larger numbers than factors, most students find them easier to work with. Read on.

Generating multiples

The preceding section, "Fabulous Factors," tells you how to find *all* the factors of a number. Finding all the factors is possible because factors of a number are always less than or equal to the number itself. So no matter how large a number is, it always has a finite (limited) number of factors.

Unlike factors, multiples of a number are greater than or equal to the number itself. (The only exception to this is 0, which is a multiple of every number.) Because of this, the multiples of a number go on forever — that is, they're infinite. Nevertheless, generating a *partial* list of multiples for any number is simple.

To list multiples of any number, write down that number and then multiply it by 2, 3, 4, and so forth.

For example, here are the first few positive multiples of 7:

7   14   21   28   35   42

As you can see, this list of multiples is simply part of the multiplication table for the number 7. (For the multiplication table up to $9 \cdot 9$, see Chapter 3.)

Finding the least common multiple (LCM)

The *least common multiple* (LCM) of a set of numbers is the lowest positive number that's a multiple of every number in that set.

For example, the LCM of the numbers 2, 3, and 5 is 30 because

✔ 30 is a multiple of 2 (2 · 15 = 30)

✔ 30 is a multiple of 3 (3 · 10 = 30)

✔ 30 is a multiple of 5 (5 · 6 = 30)

✔ No number lower than 30 is a multiple of all three numbers

In this section, I give you two ways to find the LCM of two or more numbers.

Using the multiplication table to find the LCM

To find the LCM of a set of numbers, take each number in the set and jot down a list of the first several multiples in order. The LCM is the first number that appears on every list.

When looking for the LCM of two numbers, start by listing multiples of the higher number, but stop this list when the number of multiples you've written down equals the lower number. Then start listing multiples of the lower number until one of them matches the first list.

For example, suppose you want to find the LCM of 4 and 6. Begin by listing multiples of the higher number, which is 6. In this case, list only four of these multiples because the lower number is 4.

Multiples of 6:   6, 12, 18, 24, ...

Now, start listing multiples of 4:

Multiples of 4:   4, 8, 12, ...

Because 12 is the first number to appear on both lists of multiples, 12 is the LCM of 4 and 6.

This method works especially well when you want to find the LCM of two numbers, but it may take longer if you have more numbers. When working with three numbers, first multiply the two lowest numbers. For the second-highest number, find the product of the other two numbers and list that many multiples. Repeat for the lowest number.

Suppose, for instance, you want to find the LCM of 2, 3, and 5. Again, start with the highest number — in this case, 5 — listing six numbers (the product of the other two numbers, 2 · 3 = 6):

Multiples of 5:   5, 10, 15, 20, 25, 30, ...

Next, list multiples of 3, listing ten of them (because 2 · 5 = 10):

Multiples of 3:   3, 6, 9, 12, 15, 18, 21, 24, 27, 30, ...

The only numbers repeated on both lists are 15 and 30. In this case, you can save yourself the trouble of making the last list because 30 is obviously a multiple of 2, and 15 isn't. So 30 is the LCM of 2, 3, and 5.

Using prime factorization to find the LCM

A second method for finding the LCM of a set of numbers is to use the prime factorizations of those numbers. Here's how:

1. **List the prime factors of each number.**

   I show you how to find the prime factors of a number earlier in this chapter, in "Prime factors."

   Suppose you want to find the LCM of 18 and 24. List the prime factors of each number:

   18 = 2 · 3 · 3

   24 = 2 · 2 · 2 · 3

2. **For each prime number listed, underline the most repeated occurrence of this number in any prime factorization.**

   The number 2 appears once in the prime factorization of 18 but three times in that of 24, so underline the three 2s:

   18 = 2 · 3 · 3

   24 = <u>2</u> · <u>2</u> · <u>2</u> · 3

   Similarly, the number 3 appears twice in the prime factorization of 18 but only once in that of 24, so underline the two 3s:

   18 = 2 · <u>3</u> · <u>3</u>

   24 = <u>2</u> · <u>2</u> · <u>2</u> · 3

3. **Multiply all the underlined numbers.**

   Here's the product:

   2 · 2 · 2 · 3 · 3 = 72

   So the LCM of 18 and 24 is 72. This checks out because

   18 · 4 = 72

   24 · 3 = 72

Part III

Parts of the Whole: Fractions, Decimals, and Percents

The fugitives left 200 miles from here traveling 70 miles per hour. We're traveling in the opposite direction going 60 miles per hour. At what point... hey, donut shop, donut shop! Pull over!

In this part . . .

Math represents parts of the whole as fractions, decimals, and percents. Even though they may look different, all three of these concepts are closely related to division. In these chapters, you see how to apply the Big Four operations (adding, subtracting, multiplying, and dividing) to fractions, decimals, and percents. I discuss proper and improper fractions, how to reduce and increase the terms of fractions, and how to work with mixed numbers. I also talk about terminating, non-terminating, and repeating decimals. I show you how to use the percent circle to solve three common types of percent problems. I also explain how to translate fractions, decimals, and percents into each of the other two forms.

Chapter 9

Fooling with Fractions

In This Chapter

▶ Looking at basic fractions

▶ Knowing the numerator from the denominator

▶ Understanding proper fractions, improper fractions, and mixed numbers

▶ Increasing and reducing the terms of fractions

▶ Converting between improper fractions and mixed numbers

▶ Using cross-multiplication to compare fractions

Suppose that today is your birthday and your friends are throwing you a surprise party. After opening all your presents, you finish blowing out the candles on your cake, but now you have a problem: Eight of you want some cake, but you have only *one cake*. Several solutions are proposed:

✔ You could all go into the kitchen and bake seven more cakes.

✔ Instead of eating cake, everyone could eat celery sticks.

✔ Because it's your birthday, you could eat the *whole* cake and everyone else could eat celery sticks. (That was your idea.)

✔ You could cut the cake into eight equal slices so that everyone can enjoy it.

After careful consideration, you choose the last option. With that decision, you've opened the door into the exciting world of fractions. Fractions represent parts of a thing that can be cut into pieces. In this chapter, I give you some basic information about fractions that you need to know, including the three basic types of fractions: *proper fractions*, *improper fractions*, and *mixed numbers*.

I move on to increasing and reducing the terms of fractions, which you need when you begin applying the Big Four operations to fractions in Chapter 10. I also show you how to convert between improper fractions and mixed numbers. Finally, I show you how to compare fractions using cross-multiplication. By the time you're done with this chapter, you'll see how fractions really can be a piece of cake!

Slicing a Cake into Fractions

Here's a simple fact: When you cut a cake into two equal pieces, each piece is half of the cake. As a fraction, you write that as ½. In Figure 9-1, the shaded piece is half of the cake.

Figure 9-1:
Two halves of a cake.

Every fraction is made up of two numbers separated by a line, or a fraction bar. The line can be either diagonal or horizontal — so you can write this fraction in either of the following two ways:

$$\frac{1}{2} \qquad\qquad ½$$

The number above the line is called the *numerator*. The numerator tells you how many pieces you have. In this case, you have one dark-shaded piece of cake, so the numerator is 1.

The number below the line is called the *denominator*. The denominator tells you how many equal pieces the whole cake has been cut up into. In this case, the denominator is 2.

Similarly, when you cut a cake into three equal slices, each piece is a third of the cake (see Figure 9-2).

Figure 9-2:
Cake cut into thirds.

This time, the shaded piece is one-third (⅓) of the cake. Again, the numerator tells you how many pieces you have, and the denominator tells you how many equal pieces the whole cake has been cut up into.

Figure 9-3 shows a few more examples of ways to represent parts of the whole with fractions.

A.
C.

B.
D.

Figure 9-3: Cakes cut and shaded into (A) ¾, (B) ⅚, (C) ¹⁄₁₀, and (D) ⁷⁄₁₀.

In each case, the numerator tells you how many pieces are shaded, and the denominator tells how many pieces there are altogether.

The fraction bar can also mean a division sign. In other words, ¾ signifies 3 ÷ 4. If you take three cakes and divide them among four people, each person gets ¾ of a cake.

Knowing the Fraction Facts of Life

Fractions have their own special vocabulary and a few important properties that are worth knowing right from the start. When you know them, you find working with fractions a lot easier.

Telling the numerator from the denominator

The top number in a fraction is called the *numerator,* and the bottom number is called the *denominator.* For example, look at the following fraction:

$$\frac{3}{4}$$

In this example, the number 3 is the numerator, and the number 4 is the denominator. Similarly, look at this fraction:

$$\frac{55}{89}$$

The number 55 is the numerator, and the number 89 is the denominator.

Flipping for reciprocals

When you flip a fraction over, you get its *reciprocal*. For example, the following numbers are reciprocals:

$$\frac{2}{3} \quad \text{and} \quad \frac{3}{2}$$

$$\frac{11}{14} \quad \text{and} \quad \frac{14}{11}$$

$\frac{19}{19}$ is its own reciprocal

Using ones and zeros

When the denominator (bottom number) of a fraction is 1, the fraction is equal to the numerator by itself. Or conversely, you can turn any whole number into a fraction by drawing a line and placing the number 1 under it. For example,

$$\frac{2}{1} = 2 \qquad \frac{9}{1} = 9 \qquad \frac{157}{1} = 157$$

When the numerator and denominator match, the fraction equals 1. That's because if you cut a cake into eight pieces and you keep all eight of them, you have the entire cake. Here are some fractions that equal 1:

$$\frac{8}{8} = 1 \qquad \frac{11}{11} = 1 \qquad \frac{365}{365} = 1$$

When the numerator of a fraction is 0, the fraction is equal to 0. For example,

$$\frac{0}{1} = 0 \qquad \frac{0}{12} = 0 \qquad \frac{0}{113} = 0$$

The denominator of a fraction can *never* be 0. Fractions with 0 in the denominator are *undefined* — that is, they have no mathematical meaning.

Remember from earlier in this chapter that placing a number in the denominator is similar to cutting a cake up into that number of pieces. You can cut a cake into two, or ten, or even a million pieces. You can even cut it into one piece (that is, don't cut it at all). But you can't cut a cake into *zero* pieces. For this reason, putting 0 in the denominator — much like lighting an entire book of matches on fire — is something you should never, *never* do.

Mixing things up

A mixed number is a combination of a whole number and a proper fraction added together. Here are some examples:

$$1\frac{1}{2} \qquad 5\frac{3}{4} \qquad 99\frac{44}{100}$$

A mixed number is always equal to the whole number *plus* the fraction attached to it. That is, 1½ means 1 + ½, 5¾ means 5 + ¾, and so on.

Knowing proper from improper

When the numerator and the denominator are equal, the fraction *equals* 1:

$$\frac{2}{2} = 1 \qquad \frac{5}{5} = 1 \qquad \frac{78}{78} = 1$$

When the numerator (top number) is less than the denominator (bottom number), the fraction is *less than* 1:

$$\frac{1}{2} < 1 \qquad \frac{3}{5} < 1 \qquad \frac{63}{73} < 1$$

Fractions like these are called are called *proper fractions*. Positive proper fractions are always between 0 and 1. However, when the numerator is greater than the denominator, the fraction is *greater than* 1. Take a look:

$$\frac{3}{2} > 1 \qquad \frac{7}{4} > 1 \qquad \frac{98}{97} > 1$$

Any fraction that's greater than 1 is called an *improper fraction*. It's customary to convert an improper fraction to a mixed number, especially when it's the final answer to a problem.

An improper fraction is always top heavy, as if it's unstable and wants to fall over. To stabilize it, convert it to a mixed number. On the other hand, proper fractions are always stable.

Later in this chapter, I discuss improper fractions in more detail when I show you how to convert between improper fractions and mixed numbers.

Increasing and Reducing Terms of Fractions

Take a look at these three fractions:

$$\frac{1}{2} \qquad \frac{2}{4} \qquad \frac{3}{6}$$

If you cut three cakes (as I do earlier in this chapter) into these three fractions, exactly half of the cake will be shaded just like in Figure 9-1, no matter how you slice it. (Get it? No matter how you *slice* it? You may as well laugh at the bad jokes, too — they're free.) The important thing here isn't the humor, or the lack of it, but the idea about fractions.

The fractions ½, ¾, and ⅗ are all equal in value. In fact, you can write down lots of fractions that are also equal to these. As long as the numerator is exactly half the denominator, the fractions are all equal to ½ — for example,

$$\frac{11}{22} \qquad \frac{100}{200} \qquad \frac{1,000,000}{2,000,000}$$

These fractions are equal to ½, but their *terms* (the numerator and denominator) are different. In this section, I show you how to both increase and reduce the terms of a fraction without changing its value.

Increasing the terms of fractions

To increase the terms of a fraction by a certain number, multiply both the numerator and the denominator by that number.

For example, to increase the terms of the fraction ¾ by 2, multiply both the numerator and the denominator by 2:

$$\frac{3}{4} = \frac{3 \cdot 2}{4 \cdot 2} = \frac{6}{8}$$

Similarly, to increase the terms of the fraction ⁵⁄₁₁ by 7, multiply both the numerator and the denominator by 7:

$$\frac{5}{11} = \frac{5 \cdot 7}{11 \cdot 7} = \frac{35}{77}$$

Increasing the terms of a fraction doesn't change its value. Because you're multiplying the numerator and denominator by the same number, you're essentially multiplying the fraction by a fraction that equals 1.

One key thing to know is how to increase the terms of a fraction so that the denominator becomes a preset number. Here's how you do it:

1. **Divide the new denominator by the old denominator.**

   To keep the fractions equal, you have to multiply the numerator and denominator of the old fraction by the same number. This first step tells you what the old denominator was multiplied by to get the new one.

   For example, suppose you want to raise the terms of the fraction ⁴⁄₇ so that the denominator is 35. That is, you're trying to fill in the question mark here:

   $$\frac{4}{7} = \frac{?}{35}$$

   Divide 35 by 7, which tells you that the denominator was multiplied by 5.

2. **Multiply this result by the old numerator to get the new numerator.**

   You now know how the two denominators are related. The numerators need to have the same relationship, so multiply the old numerator by the number you found in Step 1.

   Multiply 5 by 4, which gives you 20. So here's the answer:

   $$\frac{4}{7} = \frac{4 \cdot 5}{7 \cdot 5} = \frac{20}{35}$$

Reducing fractions to lowest terms

Reducing fractions is similar to increasing fractions, except it involves division rather than multiplication. But because you can't always divide, reducing takes a bit more finesse.

In practice, reducing fractions is similar to factoring numbers. For this reason, if you're not up on factoring, you may want to review this topic in Chapter 8.

In this section, I show you the formal way to reduce fractions, which works in all cases. Then I show you a more informal way that you can use after you're more comfortable.

Reducing fractions the formal way

Reducing fractions the formal way relies on an understanding of how to break down a number into its prime factors. I discuss this in detail in Chapter 8, so if you're shaky on this concept, you may want to review it first.

Here's how to reduce a fraction:

1. **Break down both the numerator (top number) and denominator (bottom number) into their prime factors.**

   For example, suppose you want to reduce the fraction $\frac{12}{30}$. Break down both 12 and 30 into their prime factors:

   $$\frac{12}{30} = \frac{2 \cdot 2 \cdot 3}{2 \cdot 3 \cdot 5}$$

2. **Cross out any common factors.**

   As you can see, I cross out a 2 and a 3, because they're common factors — that is, they appear in both the numerator and denominator:

   $$\frac{12}{30} = \frac{2 \cdot 2 \cdot 3}{2 \cdot 3 \cdot 5}$$

3. **Multiply the remaining numbers to get the reduced numerator and denominator.**

   This shows you that the fraction $\frac{12}{30}$ reduces to $\frac{2}{5}$:

   $$\frac{12}{30} = \frac{2 \cdot 2 \cdot 3}{2 \cdot 3 \cdot 5} = \frac{2}{5}$$

As another example, here's how you reduce the fraction $\frac{32}{100}$:

$$\frac{32}{100} = \frac{2 \cdot 2 \cdot 2 \cdot 2 \cdot 2}{2 \cdot 2 \cdot 5 \cdot 5} = \frac{8}{25}$$

This time, cross out two 2s from both the top and the bottom as common factors. The remaining 2s on top, and the 5s on the bottom, aren't common factors. So the fraction $\frac{32}{100}$ reduces to $\frac{8}{25}$.

Reducing fractions the informal way

Here's an easier way to reduce fractions after you get comfortable with the concept:

1. **If the numerator (top number) and denominator (bottom number) are both divisible by 2 — that is, if they're both even — divide both by 2.**

   For example, suppose you want to reduce the fraction $\frac{24}{60}$. The numerator and the denominator are both even, so divide them both by 2:

   $$\frac{24}{60} = \frac{12}{30}$$

2. **Repeat Step 1 until the numerator or denominator (or both) is no longer divisible by 2.**

   In the resulting fraction, both numbers are still even, so repeat the first step again:

   $$\frac{12}{30} = \frac{6}{15}$$

3. **Repeat Step 1 using the number 3, and then 5, and then 7, continuing testing prime numbers until you're sure that the numerator and denominator have no common factors.**

   Now, the numerator and the denominator are both divisible by 3 (see Chapter 7 for easy ways to tell if one number is divisible by another), so divide both by 3:

   $$\frac{6}{15} = \frac{2}{5}$$

   Neither the numerator nor the denominator is divisible by 3, so this step is complete. At this point, you can move on to test for divisibility by 5, 7, and so on, but you really don't need to. The numerator is 2, and it obviously isn't divisible by any larger number, so you know that the fraction $\frac{24}{60}$ reduces to $\frac{2}{5}$.

Converting Between Improper Fractions and Mixed Numbers

In "Knowing the fraction facts of life," I tell you that any fraction whose numerator is greater than its denominator is an *improper fraction*. Improper fractions are very useful and easy to work with, but for some reason, people just don't like them. (The word *improper* should've tipped you off.) Teachers especially don't like them, and they really don't like an improper fraction to appear as the answer to a problem. However, they love mixed numbers. One reason they love them is that estimating the approximate size of a mixed number is easy.

For example, if I tell you to put $\frac{31}{3}$ of a gallon of gasoline in my car, you'd probably find it hard to estimate roughly how much that is: 5 gallons, 10 gallons, 20 gallons?

But if I tell you to get $10\frac{1}{3}$ gallons of gasoline, you know immediately that this amount is a little more than 10 but less than 11 gallons. Although $10\frac{1}{3}$ is the same as $\frac{31}{3}$, knowing the mixed number is a lot more helpful in practice. For this reason, you often have to convert improper fractions to mixed numbers.

Knowing the parts of a mixed number

Every mixed number has both a whole number part and a fractional part. So the three numbers in a mixed number are

- The whole number
- The numerator
- The denominator

For example, in the mixed number 3½, the whole number part is 3 and the fractional part is ½. So this mixed number is made up of three numbers: the whole number (3), the numerator (1), and the denominator (2). Knowing these three parts of a mixed number is helpful for converting back and forth between mixed numbers and improper fractions.

Converting a mixed number to an improper fraction

To convert a mixed number to an improper fraction,

1. **Multiply the denominator of the fractional part by the whole number, and add the result to the numerator.**

   For example, suppose you want to convert the mixed number 5⅔ to an improper fraction. First, multiply 3 by 5 and add 2:

   $$(3 \cdot 5) + 2 = 17$$

2. **Use this result as your numerator, and place it over the denominator you already have.**

   Place this result over the denominator:

   $$\frac{17}{3}$$

   So the mixed number 5⅔ equals the improper fraction ¹⁷⁄₃. This method works for all mixed numbers. Furthermore, if you start with the fractional part reduced, the answer will also be reduced (see the earlier "Increasing and Reducing Terms of Fractions" section).

Converting an improper fraction to a mixed number

To convert an improper fraction to a mixed number, divide the numerator by the denominator (see Chapter 3). Then write the mixed number in this way:

- ✔ The quotient (answer) is the whole-number part.
- ✔ The remainder is the numerator.
- ✔ The denominator of the improper fraction is the denominator.

For example, suppose you want to write the improper fraction $\frac{19}{5}$ as a mixed number. First, divide 19 by 5:

$$19 \div 5 = 3 \text{ r } 4$$

Then write the mixed number as follows:

$$3 \frac{4}{5}$$

This method works for all improper fractions. And, as is true of conversions in the other direction, if you start with a reduced fraction, you don't have to reduce your answer (see "Increasing and reducing terms of fractions").

Understanding Cross-multiplication

Cross-multiplication is a handy little thing to know. You can use it in a few different ways, so I explain it here, and then I show you an immediate application.

To cross-multiply two fractions,

1. **Multiply the numerator of the first fraction by the denominator of the second fraction and jot down the answer.**

2. **Multiply the numerator of the second fraction by the denominator of the first fraction and jot down the answer.**

For example, suppose you have these two fractions:

$$\frac{2}{9} \bowtie \frac{4}{7}$$

When you cross-multiply, you get these two numbers:

$$2 \cdot 7 = 14 \qquad 4 \cdot 9 = 36$$

You can use cross-multiplication to compare fractions and find out which is greater. When you do so, make sure that you *start* with the numerator of the first fraction.

To find out which of two fractions is larger, cross-multiply and place the two numbers you get, in order, under the two fractions. The larger number is always under the larger fraction.

For example, suppose you want to find out which of the following three fractions is the greatest:

$$\frac{3}{5} \qquad \frac{5}{9} \qquad \frac{6}{11}$$

Cross-multiplication works only with two fractions at a time, so pick the first two:

$$\frac{3}{5} \bowtie \frac{5}{9}$$
$$3 \cdot 9 = 27 \qquad 5 \cdot 5 = 25$$

Because 27 is greater than 25, you know now that ⅗ is greater than ⁵⁄₉. So you can throw out ⁵⁄₉.

Now do the same thing for ⅗ and ⁶⁄₁₁:

$$\frac{3}{5} \bowtie \frac{6}{11}$$
$$3 \cdot 11 = 33 \qquad 6 \cdot 5 = 30$$

Because 33 is greater than 30, ⅗ is greater than ⁶⁄₁₁. Pretty straightforward, right? And that's all you have to know for now. I show you a bunch of great things you can do with this simple skill in the next chapter.

Chapter 10

Parting Ways: Fractions and the Big Four Operations

In this chapter, the focus is on applying the Big Four operations to fractions. I start out by showing you how to multiply and divide fractions, which isn't much more difficult than multiplying whole numbers. Surprisingly, adding and subtracting fractions is a bit trickier. I show you a variety of methods, each with its own strengths and weaknesses, and I recommend how to choose which method will work best depending on the problem you have to solve.

Later in the chapter, I move on to mixed numbers. Again, multiplication and division shouldn't pose too much of a problem because the process in each case is almost the same as multiplying and dividing fractions. I save adding and subtracting mixed numbers for the very end. By then, you should be much more comfortable with fractions and ready to tackle the challenge.

Multiplying and Dividing Fractions

One of the odd little ironies of life is that multiplying and dividing fractions is easier than adding or subtracting them — just two easy steps and you're done! For this reason, I show you how to multiply and divide fractions before I show you how to add or subtract them. In fact, you may find multiplying fractions easier than multiplying whole numbers because the numbers you're working with are usually small. More good news is that dividing fractions is nearly as easy as multiplying them. So I'm not even wishing you good luck — you don't need it!

Multiplying numerators and denominators straight across

Everything in life should be as simple as multiplying fractions. All you need for multiplying fractions is a pen or pencil, something to write on (preferably not your hand), and a basic knowledge of the multiplication table. (See Chapter 3 for a multiplication refresher.)

Here's how to multiply two fractions:

1. **Multiply the numerators (the numbers on top) together to get the numerator of the answer.**

2. **Multiply the denominators (the numbers on the bottom) together to get the denominator of the answer.**

For example, here's how to multiply ⅖ · 3/7:

$$\frac{2}{5} \cdot \frac{3}{7} = \frac{2 \cdot 3}{5 \cdot 7} = \frac{6}{35}$$

Sometimes, when you multiply fractions, you may have an opportunity to reduce to lowest terms. (For more on when and how to reduce a fraction, see Chapter 9.) As a rule, math people are crazy about reduced fractions, and teachers sometimes take points off a right answer if you could've reduced it but didn't. Here's a multiplication problem that ends up with an answer that's not in its lowest terms:

$$\frac{4}{5} \cdot \frac{7}{8} = \frac{4 \cdot 7}{5 \cdot 8} = \frac{28}{40}$$

Because the numerator and the denominator are both even numbers, this fraction can be reduced. Start by dividing both numbers by 2:

$$\frac{28 \div 2}{40 \div 2} = \frac{14}{20}$$

Again, the numerator and the denominator are both even, so do it again:

$$\frac{14 \div 2}{20 \div 2} = \frac{7}{10}$$

This fraction is now fully reduced.

When multiplying fractions, you can often make your job easier by canceling out equal factors in the numerator and denominator. Canceling out equal factors makes the numbers that you're multiplying smaller and easier to work with, and it also saves you the trouble of reducing at the end. Here's how it works:

✔ When the numerator of one fraction and the denominator of the other are the same, change both of these numbers to 1s. (See the nearby sidebar for why this works.)

✔ When the numerator of one fraction and the denominator of the other are divisible by the same number, factor this number out of both. In other words, divide the numerator and denominator by that common factor. (For more on how to find factors, see Chapter 8.)

For example, suppose you want to multiply the following two numbers:

$$\frac{5}{13} \cdot \frac{13}{20}$$

You can make this problem easier by canceling out the number 13 as follows:

$$\frac{5}{1 \, \cancel{13}} \cdot \frac{\cancel{13} \, 1}{20} = \frac{5 \cdot 1}{1 \cdot 20} = \frac{5}{20}$$

You can make it even easier by noticing that $20 = 5 \cdot 4$, so you can factor out the number 5 as follows:

$$\frac{1 \, \cancel{5}}{1} \cdot \frac{1}{\cancel{20} \, 4} = \frac{1 \cdot 1}{1 \cdot 4} = \frac{1}{4}$$

One is the easiest number

With fractions, the relationship between the numbers, not the actual numbers themselves, is most important. Understanding how to multiply and divide fractions can give you a deeper understanding of why you can increase or decrease the numbers within a fraction without changing the value of the whole fraction.

When you multiply or divide any number by 1, the answer is the same number. This rule also goes for fractions, so

$$\frac{3}{8} \cdot 1 = \frac{3}{8} \text{ and } \frac{3}{8} \div 1 = \frac{3}{8}$$

$$\frac{5}{13} \cdot 1 = \frac{5}{13} \text{ and } \frac{5}{13} \div 1 = \frac{5}{13}$$

$$\frac{67}{70} \cdot 1 = \frac{67}{70} \text{ and } \frac{67}{70} \div 1 = \frac{67}{70}$$

And as I discuss in Chapter 9, when a fraction has the same number in both the numerator and the denominator, its value is 1. In other words, the fractions ²⁄₂, ³⁄₃, and ⁴⁄₄ are all equal to 1. Look what happens you multiply the fraction ¾ by ²⁄₂:

$$\frac{3}{4} \cdot \frac{2}{2} = \frac{3 \cdot 2}{4 \cdot 2} = \frac{6}{8}$$

The net effect is that you've increased the terms of the original fraction by 2. But all you've done is multiply the fraction by 1, so the value of the fraction hasn't changed. The fraction ⁶⁄₈ is equal to ¾.

Similarly, reducing the fraction ⁶⁄₉ by a factor of 3 is the same as dividing that fraction by ³⁄₃ (which is equal to 1):

$$\frac{6}{9} \div \frac{3}{3} = \frac{6 \div 3}{9 \div 3} = \frac{2}{3}$$

So ⁶⁄₉ is equal to ²⁄₃.

Doing a flip to divide fractions

Dividing fractions is just as easy as multiplying them. In fact, when you divide fractions, you really turn the problem into multiplication.

To divide one fraction by another, multiply the first fraction by the reciprocal of the second. (As I discuss in Chapter 9, the *reciprocal* of a fraction is simply that fraction turned upside down.)

For example, here's how you turn fraction division into multiplication:

$$\frac{1}{3} \div \frac{4}{5} = \frac{1}{3} \cdot \frac{5}{4}$$

As you can see, I turn ⅘ into its reciprocal — ¾ — and change the division sign to a multiplication sign. After that, just multiply the fractions as I describe in "Multiplying numerators and denominators straight across":

$$\frac{1}{3} \cdot \frac{5}{4} = \frac{1 \cdot 5}{3 \cdot 4} = \frac{5}{12}$$

As with multiplication, in some cases you may have to reduce your result at the end. But you can also make your job easier by canceling out equal factors. (See the preceding section.)

All Together Now: Adding Fractions

When you add fractions, one important thing to notice is whether their denominators (the numbers on the bottom) are the same. If they're the same — woohoo! Adding fractions that have the same denominator is a walk in the park. But when fractions have different denominators, adding them becomes a tad more complex.

To make matters worse, many teachers make adding fractions even more difficult by requiring you to use a long and complicated method when, in many cases, a short and easy one will do.

In this section, I first show you how to add fractions with the same denominator. Then I show you a foolproof method for adding fractions when the denominators are different. It always works, and it's usually the simplest way to go. After that, I show you a quick method that you can use only for certain problems. Finally, I show you the longer, more complicated way to add fractions that usually gets taught.

Finding the sum of fractions with the same denominator

To add two fractions that have the same denominator (bottom number), add the numerators (top numbers) together and leave the denominator unchanged.

For example, consider the following problem:

$$\frac{1}{5} + \frac{2}{5} = \frac{1+2}{5} = \frac{3}{5}$$

As you can see, to add these two fractions, you add the numerators (1 + 2) and keep the denominator (5).

Why does this work? Chapter 9 tells you that you can think about fractions as pieces of cake. The denominator in this case tells you that the entire cake has been cut into five pieces. So when you add ⅕ + ⅖, you're really adding one piece plus two pieces. The answer, of course, is three pieces — that is, ⅗.

Even if you have to add more than two fractions, as long as the denominators are all the same, you just add the numerators and leave the denominator unchanged:

$$\frac{1}{17} + \frac{3}{17} + \frac{4}{17} + \frac{6}{17} = \frac{1+3+4+6}{17} = \frac{14}{17}$$

Sometimes, when you add fractions with the same denominator, you may have to reduce it to lowest terms (to find out more about reducing, flip to Chapter 9). Take this problem for example:

$$\frac{1}{4} + \frac{1}{4} = \frac{1+1}{4} = \frac{2}{4}$$

The numerator and the denominator are both even, so you know they can be reduced:

$$\frac{2}{4} = \frac{1}{2}$$

In other cases, the sum of two proper fractions is an improper fraction. You get a numerator that's larger than the denominator when the two fractions add up to more than 1, as in this case:

$$\frac{3}{7} + \frac{5}{7} = \frac{8}{7}$$

If you have more work to do with this fraction, leave it as an improper fraction so that it's easier to work with. But if this is your final answer, you may need to turn it into a mixed number (I cover mixed numbers in Chapter 9):

$$\frac{8}{7} = 8 \div 7 = 1\,r\,1 = 1\frac{1}{7}$$

When two fractions have the same numerator, don't add them by adding the denominators and leaving the numerator unchanged.

Adding fractions with different denominators

When the fractions that you want to add have different denominators, adding them isn't quite as easy. At the same time, it doesn't have to be as hard as most teachers make it.

Now, I'm shimmying out onto a brittle limb here, but this needs to be said: There's a very simple way to add fractions. It always works. It makes adding fractions only a little more difficult than multiplying them. And as you move up the math food chain into algebra, it becomes the most useful method.

So why doesn't anybody talk about it? I think it's a clear case of tradition being stronger than common sense. The traditional way to add fractions is more difficult, more time-consuming, and more likely to cause an error. But generation after generation has been taught that it's the right way to add fractions. It's a vicious cycle.

But in this book, I'm breaking with tradition. I first show you the easy way to add fractions. Then I show you a quick trick that works in a few special cases. And finally, I show you the traditional way to add fractions.

Using the easy way

At some point in your life, I bet some teacher somewhere told you these golden words of wisdom: "You can't add two fractions with different denominators." Your teacher was wrong! Here's the way to do it:

1. **Cross-multiply the two fractions and add the results together to get the numerator of the answer.**

   Suppose you want to add the fractions ⅓ and ⅖. To get the numerator of the answer, cross-multiply. In other words, multiply the numerator of each fraction by the denominator of the other:

   $$\frac{1}{3} + \frac{2}{5}$$

   $$1 \cdot 5 = 5$$

   $$2 \cdot 3 = 6$$

   Add the results to get the numerator of the answer:

   $$5 + 6 = 11$$

2. **Multiply the two denominators together to get the denominator of the answer.**

   To get the denominator, just multiply the denominators of the two fractions:

   $$3 \cdot 5 = 15$$

   The denominator of the answer is 15.

3. **Write your answer as a fraction.**

   $$\frac{1}{3} + \frac{2}{5} = \frac{11}{15}$$

As you discover in the earlier section "Finding the sum of fractions with the same denominator," when you add fractions, you sometimes need to reduce the answer that you get. Here's an example:

$$\frac{5}{8} + \frac{3}{10} = \frac{5 \cdot 10 + 3 \cdot 8}{8 \cdot 10} = \frac{50 + 24}{80} = \frac{74}{80}$$

Because the numerator and the denominator are both even numbers, you know that the fraction can be reduced. So try dividing both numbers by 2:

$$\frac{74 \div 2}{80 \div 2} = \frac{37}{40}$$

This fraction can't be reduced further, so ³⁷⁄₄₀ is the final answer.

As you also discover in "Finding the sum of fractions with the same denominator," sometimes when you add two proper fractions, your answer is an improper fraction:

$$\frac{4}{5} + \frac{3}{7} = \frac{4 \cdot 7 + 3 \cdot 5}{5 \cdot 7} = \frac{28 + 15}{35} = \frac{43}{35}$$

If you have more work to do with this fraction, leave it as an improper fraction so that it's easier to work with. But if this is your final answer, you may need to turn it into a mixed number (see Chapter 9 for details).

$$\frac{43}{35} = 43 \div 35 = 1\, r\, 8 = 1\frac{8}{35}$$

In some cases, you may have to add more than one fraction. The method is similar, with one small tweak. For example, suppose you want to add ½ + ⅗ + ⁴⁄₇:

1. **Start out by multiplying the *numerator* of the first fraction by the *denominators* of all the other fractions.**

   $$\frac{1}{2} + \frac{3}{5} + \frac{4}{7}$$
   $$(1 \cdot 5 \cdot 7) = 35$$

2. **Do the same with the second fraction and add this value to the first.**

$$\frac{1}{2} + \frac{3}{5} + \frac{4}{7}$$

$$35 + (3 \cdot 2 \cdot 7) = 35 + 42$$

3. **Do the same with the remaining fraction(s).**

$$\frac{1}{2} + \frac{3}{5} + \frac{4}{7}$$

$$35 + 42 + (4 \cdot 2 \cdot 5) = 35 + 42 + 40 = 117$$

When you're done, you have the numerator of the answer.

4. **To get the denominator, just multiply all the denominators together:**

$$\frac{1}{2} + \frac{3}{5} + \frac{4}{7}$$

$$= \frac{35 + 42 + 40}{2 \cdot 5 \cdot 7} = \frac{117}{70}$$

As usual, you may need to reduce or change an improper fraction to a mixed number. In this example, you just need to change to a mixed number (see Chapter 9 for details):

$$\frac{117}{70} = 117 \div 70 = 1\,\mathrm{r}\,47 = 1\frac{47}{70}$$

Trying a quick trick

I show you a way to add fractions with different denominators in the preceding section. It always works and it's easy. So why do I want to show you another way? It feels like déjà vu.

In some cases, you can save yourself a lot of effort with a little bit of smart thinking. You can't always use this method, but you can use it when one denominator is a multiple of the other. (For more on multiples, see Chapter 8.) Look at the following problem:

$$\frac{11}{12} + \frac{19}{24}$$

First, I solve it the way I show you in the preceding section:

$$\frac{11}{12} + \frac{19}{24} = \frac{11 \cdot 24 + 19 \cdot 12}{12 \cdot 24} = \frac{264 + 228}{288} = \frac{492}{288}$$

Those are some big numbers, and I'm still not done because the numerator is larger than the denominator. The answer is an improper fraction. Worse yet, the numerator and denominator are both even numbers, so the answer still needs to be reduced.

With certain fraction addition problems, I can give you a smarter way to work. The trick is to turn a problem with different denominators into a much easier problem with the same denominator.

Before you add two fractions with different denominators, check the denominators to see whether one is a multiple of the other (for more on multiples, flip to Chapter 8). If it is, you can use the quick trick:

1. **Increase the terms of the fraction with the smaller denominator so that it has the larger denominator.**

   Look at the earlier problem in this new way:

   $$\frac{11}{12} + \frac{19}{24}$$

   As you can see, 12 divides into 24 without a remainder. In this case, you want to raise the terms of $\frac{11}{12}$ so that the denominator is 24:

   $$\frac{11}{12} = \frac{?}{24}$$

   I show you how to do this kind of problem in Chapter 9. To fill in the question mark, the trick is to divide 24 by 12 to find out how the denominators are related; then multiply the result by 11:

   $$? = (24 \div 12) \cdot 11 = 22$$

   So $\frac{11}{12} = \frac{22}{24}$.

2. **Rewrite the problem, substituting this increased version of the fraction, and add as I show you earlier in "Finding the sum of fractions with the same denominator."**

   Now you can rewrite the problem this way:

   $$\frac{22}{24} + \frac{19}{24} = \frac{41}{24}$$

   As you can see, the numbers in this case are much smaller and easier to work with. The answer here is an improper fraction; changing it into a mixed number is easy:

   $$\frac{41}{24} = 41 \div 24 = 1\,r\,17 = 1\frac{17}{24}$$

Relying on the traditional way

In the two preceding sections, I show you two ways to add fractions with different denominators. They both work great depending on the circumstances. So why do I want to show you yet a third way? It feels like déjà vu all over again.

The truth is that I don't want to show you this way. But they're *forcing* me to. And you know who *they* are, don't you? The man — the system — the powers that be. The ones who want to keep you down in the mud, groveling at their feet. Okay, so I'm exaggerating a little. But let me impress on you that you don't have to add fractions this way unless you really want to (or unless your teacher insists on it).

Here's the traditional way to add fractions with two different denominators:

1. **Find the least common multiple (LCM) of the two denominators (for more on finding the LCM of two numbers, see Chapter 8).**

   Suppose you want to add the fractions ¾ + ⁷⁄₁₀. First find the LCM of the two denominators, 4 and 10. Here's how to find the LCM using the multiplication table method:

   - **Multiples of 10:** 10, 20, 30, 40

   - **Multiples of 4:** 4, 8, 12, 16, 20

   So the LCM of 4 and 10 is 20.

2. **Increase the terms of each fraction so that the denominator of each equals the LCM (for more on how to do this, see Chapter 9).**

   Increase each fraction to higher terms so that the denominator of each is 20.

   $$\frac{3}{4} = \frac{3 \cdot 5}{4 \cdot 5} = \frac{15}{20} \quad \text{and} \quad \frac{7}{10} = \frac{7 \cdot 2}{10 \cdot 2} = \frac{14}{20}$$

3. **Substitute these two new fractions for the original ones and add as I show you earlier in "Finding the sum of fractions with the same denominator."**

   At this point, you have two fractions that have the same denominator:

   $$\frac{15}{20} + \frac{14}{20} = \frac{29}{20}$$

   When the answer is an improper fraction, you still need to change it to a mixed number:

   $$\frac{29}{20} = 29 \div 20 = 1\,r\,9 = 1\frac{9}{20}$$

As another example, suppose you want to add the numbers ⅚ + ³⁄₁₀ + ²⁄₁₅.

1. **Find the LCM of 6, 10, and 15.**

   This time, I use the prime factorization method (see Chapter 8 for details on how to do this). Start by decomposing the three denominators down to their prime factors:

   $$6 = \underline{2} \cdot \underline{3}$$
   $$10 = 2 \cdot \underline{5}$$
   $$15 = 3 \cdot 5$$

   These denominators have a total of three different prime factors — 2, 3, and 5. Each prime factor appears only once in any decomposition, so the LCM of 6, 10, and 15 is

   $$2 \cdot 3 \cdot 5 = 30$$

2. **You need to increase the terms of all three fractions so that their denominators are 30:**

$$\frac{5}{6} = \frac{5 \cdot 5}{6 \cdot 5} = \frac{25}{30}$$

$$\frac{3}{10} = \frac{3 \cdot 3}{10 \cdot 3} = \frac{9}{30}$$

$$\frac{2}{15} = \frac{2 \cdot 2}{15 \cdot 2} = \frac{4}{30}$$

3. **Simply add the three new fractions:**

$$\frac{25}{30} + \frac{9}{30} + \frac{4}{30} = \frac{38}{30}$$

Again, you need to change this improper fraction to a mixed number:

$$\frac{38}{30} = 38 \div 30 = 1\,r\,8 = 1\frac{8}{30}$$

Because both numbers are divisible by 2, you can reduce the fraction:

$$1\frac{8}{30} = 1\frac{4}{15}$$

Pick your trick: Choosing the best method

As I say earlier in this chapter, I think that the traditional way to add fractions is more difficult than either the easy way or the quick trick. Your teacher may require you to use the traditional way, and after you get the hang of it, you'll get good at it. But given the choice, here's my recommendation:

- ✔ Use the easy way when the numerators and denominators are small (say, 15 or under).

- ✔ Use the quick trick with larger numerators and denominators when one denominator is a multiple of the other.

- ✔ Use the traditional way only when you can't use either of the other methods (or when you know the LCM just by looking at the denominators).

Take It Away: Subtracting Fractions

Subtracting fractions isn't really much different from adding them. As with addition, when the denominators are the same, subtraction is easy. And when the denominators are different, the methods I show you for adding fractions can be tweaked for subtracting them.

So to figure out how to subtract fractions, you can read the section "All Together Now: Adding Fractions" and substitute a minus sign (–) for every plus sign (+). But it'd be just a little chintzy if I expected you to do that. So in this section, I show you four ways to subtract fractions that mirror what I discuss earlier in this chapter about adding them.

Subtracting fractions with the same denominator

As with addition, subtracting fractions with the same denominator is always easy. When the denominators are the same, you can just think of the fractions as pieces of cake.

To subtract one fraction from another when they both have the same denominator (bottom number), subtract the numerator (top number) of the second from the numerator of the first and keep the denominator the same. For example:

$$\frac{3}{5} - \frac{2}{5} = \frac{3-2}{5} = \frac{1}{5}$$

Sometimes, as when you add fractions, you may have to reduce:

$$\frac{3}{10} - \frac{1}{10} = \frac{3-1}{10} = \frac{2}{10}$$

Because the numerator and denominator are both even, you can reduce this fraction by a factor of 2:

$$\frac{2}{10} = \frac{2 \div 2}{10 \div 2} = \frac{1}{5}$$

Unlike addition, when you subtract one proper fraction from another, you never get an improper fraction.

Subtracting fractions with different denominators

Just as with addition, you have a choice of methods when subtracting fractions. These three methods are similar to the methods I show you for adding fractions: the easy way, the quick trick, and the traditional way.

The easy way always works, and I recommend this method for most of your fraction subtracting needs. The quick trick is a great timesaver, so use it when you can. And as for the traditional way — well, even if I don't like it, your teacher and other math purists probably do.

Knowing the easy way

This way of subtracting fractions works in all cases, and it's easy. (In the next section, I show you a quick way to subtract fractions when one denominator is a multiple of the other.) Here's the easy way to subtract fractions that have different denominators:

1. **Cross-multiply the two fractions and subtract the second number from the first to get the numerator of the answer.**

   For example, suppose you want to subtract $\frac{6}{7} - \frac{2}{5}$. To get the numerator, cross-multiply the two fractions and then subtract the second number from the first number (see Chapter 9 for info on cross-multiplication):

   $$\frac{6}{7} - \frac{2}{5}$$
   $$(6 \cdot 5) - (2 \cdot 7) = 30 - 14 = 16$$

   After you cross-multiply, be sure to subtract in the correct order. (The first number is the numerator of the first fraction times the denominator of the second.)

2. **Multiply the two denominators together to get the denominator of the answer.**

   $$7 \cdot 5 = 35$$

3. **Putting the numerator over the denominator gives you your answer.**

   $$\frac{16}{35}$$

Here's another example to work with:

$$\frac{9}{10} - \frac{5}{6}$$

This time, I put all the steps together:

$$\frac{9}{10} - \frac{5}{6} = \frac{9 \cdot 6 - 5 \cdot 10}{10 \cdot 6}$$

With the problem set up like this, you just have to simplify the result:

$$= \frac{54 - 50}{60} = \frac{4}{60}$$

In this case, you can reduce the fraction:

$$\frac{4}{60} = \frac{1}{15}$$

Cutting it short with a quick trick

The easy way I show you in the preceding section works best when the numerators and denominators are small. When they're larger, you may be able to take a shortcut.

Before you subtract fractions with different denominators, check the denominators to see whether one is a multiple of the other (for more on multiples, see Chapter 8). If it is, you can use the quick trick:

1. **Increase the terms of the fraction with the smaller denominator so that it has the larger denominator.**

   For example, suppose you want to find $\frac{17}{20} - \frac{31}{80}$. If you cross-multiply these fractions, your results are going to be much bigger than you want to work with. But fortunately, 80 is a multiple of 20, so you can use the quick way.

   First, increase the terms of $\frac{17}{20}$ so that the denominator is 80 (for more on increasing the terms of fractions, see Chapter 9):

   $$\frac{17}{20} = \frac{?}{80}$$
   $$? = 80 \div 20 \cdot 17 = 68$$

   So $\frac{17}{20} = \frac{68}{80}$.

2. **Rewrite the problem, substituting this increased version of the fraction, and subtract as I show you earlier in "Subtracting fractions with the same denominator."**

   Here's the problem as a subtraction of fractions with the same denominator, which is much easier to solve:

   $$\frac{68}{80} - \frac{31}{80} = \frac{37}{80}$$

   In this case, you don't have to reduce to lowest terms, although in other problems you may have to. (See Chapter 9 for more on reducing fractions.)

Keeping your teacher happy with the traditional way

As I describe earlier in this chapter in "All Together Now: Adding Fractions," you should use the traditional way only as a last resort. I recommend that you use it only when the numerator and denominator are too large to use the easy way and when you can't use the quick trick.

To use the traditional way to subtract fractions with two different denominators, follow these steps:

1. **Find the least common multiple (LCM) of the two denominators (for more on finding the LCM of two numbers, see Chapter 8).**

   For example, suppose you want to subtract $\frac{7}{8} - \frac{11}{14}$. Here's how to find the LCM of 8 and 14 using the prime factorization method:

   $$8 = \underline{2} \cdot \underline{2} \cdot \underline{2}$$
   $$14 = 2 \cdot \underline{7}$$

   I underline the case where each prime factor appears most often: 2 appears three times, and 7 appears once. So the LCM of 8 and 14 is

   $$2 \cdot 2 \cdot 2 \cdot 7 = 56$$

2. **Increase each fraction to higher terms so that the denominator of each equals the LCM (for more on how to do this, see Chapter 9).**

   The denominators of both should be 56:

   $$\frac{7}{8} = \frac{7 \cdot 7}{8 \cdot 7} = \frac{49}{56}$$

   $$\frac{11}{14} = \frac{11 \cdot 4}{14 \cdot 4} = \frac{44}{56}$$

3. **Substitute these two new fractions for the original ones and subtract as I show you earlier in "Subtracting fractions with the same denominator."**

   $$\frac{49}{56} - \frac{44}{56} = \frac{5}{56}$$

   This time, you don't need to reduce, because 5 is a prime number and 56 isn't divisible by 5. In some cases, however, you have to reduce the answer to lowest terms.

Working Properly with Mixed Numbers

All the methods I describe earlier in this chapter work for both proper and improper fractions. Unfortunately, mixed numbers are ornery little critters, and you need to figure out how to deal with them on their own terms. (For more on mixed numbers, flip to Chapter 9.)

Multiplying and dividing mixed numbers

I can't give you a direct method for multiplying and dividing mixed numbers. The only way is to convert the mixed numbers to improper fractions and multiply or divide as usual. Here's how to multiply or divide mixed numbers:

1. **Convert all mixed numbers to improper fractions (see Chapter 9 for details).**

   For example, suppose you want to multiply $1\frac{3}{5} \cdot 2\frac{1}{3}$. First convert $1\frac{3}{5}$ and $2\frac{1}{3}$ to improper fractions:

   $$1\frac{3}{5} = \frac{5 \cdot 1 + 3}{5} = \frac{8}{5}$$

   $$2\frac{1}{3} = \frac{3 \cdot 2 + 1}{3} = \frac{7}{3}$$

2. **Multiply these improper fractions (as I show you earlier in this chapter, in "Multiplying and Dividing Fractions).**

$$\frac{8}{5} \cdot \frac{7}{3} = \frac{8 \cdot 7}{5 \cdot 3} = \frac{56}{15}$$

3. **If the answer is an improper fraction, convert it back to a mixed number (see Chapter 9).**

$$\frac{56}{15} = 56 \div 15 = 3\,r\,11 = 3\frac{11}{15}$$

In this case, the answer is already in lowest terms, so you don't have to reduce it.

As a second example, suppose you want to divide 3⅔ by 1⁴⁄₇.

1. **Convert 3⅔ and 1⁴⁄₇ to improper fractions:**

$$3\frac{2}{3} = \frac{3 \cdot 3 + 2}{3} = \frac{11}{3}$$

$$1\frac{4}{7} = \frac{7 \cdot 1 + 4}{7} = \frac{11}{7}$$

2. **Divide these improper fractions.**

Divide fractions by multiplying the first fraction by the reciprocal of the second (see the earlier "Multiplying and Dividing Fractions" section):

$$\frac{11}{3} \div \frac{11}{7} = \frac{11}{3} \cdot \frac{7}{11}$$

In this case, before you multiply, you can cancel out factors of 11 in the numerator and denominator:

$$\frac{1\cancel{11}}{3} \cdot \frac{7}{\cancel{11}1} = \frac{1 \cdot 7}{3 \cdot 1} = \frac{7}{3}$$

3. **Convert the answer to a mixed number.**

$$\frac{7}{3} = 7 \div 3 = 2\,r\,1 = 2\frac{1}{3}$$

Adding and subtracting mixed numbers

One way to add and subtract mixed numbers is to convert them to improper fractions, much as I describe earlier in this chapter in "Multiplying and dividing mixed numbers," and then to add or subtract them using a method from the "All Together Now: Adding Fractions" or "Take It Away: Subtracting Fractions" section. Doing so is a perfectly valid way of getting the right answer without learning a new method.

Unfortunately, teachers just love to make people add and subtract mixed numbers in their own special way. The good news is that a lot of folks find this way easier than all the converting stuff.

In pairs: Adding two mixed numbers

Adding mixed numbers looks a lot like adding whole numbers: You stack them up one on top of the other, draw a line, and add. For this reason, some students feel more comfortable adding mixed numbers than adding fractions. Here's how to add two mixed numbers:

1. **Add the fractional parts using any method you like, and if necessary, change this sum to a mixed number and reduce it.**

2. **If the answer you found in Step 1 is an improper fraction, change it to a mixed number, write down the fractional part, and *carry* the whole number part to the whole number column.**

3. **Add the whole number parts (including any number carried).**

Your answer may also need to be reduced to lowest terms (see Chapter 9). In the examples that follow, I show you everything you need to know.

Summing up mixed numbers when the denominators are the same

As with any problem involving fractions, adding is always easier when the denominators are the same. For example, suppose you want to add $3\frac{1}{3} + 5\frac{1}{3}$. Doing mixed number problems is often easier if you place one number above the other:

$$3\frac{1}{3}$$
$$+\ 5\frac{1}{3}$$

As you can see, this arrangement is similar to how you add whole numbers, but it includes an extra column for fractions. Here's how you add these two mixed numbers step by step:

1. **Add the fractions.**

$$\frac{1}{3} + \frac{1}{3} = \frac{2}{3}$$

2. **Switch improper fractions to mixed numbers; write down your answer.**

Because $\frac{2}{3}$ is a proper fraction, you don't have to change it.

3. **Add the whole number parts.**

$$3 + 5 = 8$$

Here's how your problem looks in column form:

$$3\frac{1}{3}$$
$$+\ 5\frac{1}{3}$$
$$\overline{8\frac{2}{3}}$$

This problem is about as simple as they get. In this case, all three steps are pretty easy. But sometimes, Step 2 requires more attention. For example, suppose you want to add 8⅗ + 6⅘. Here's how you do it:

1. **Add the fractions.**

$$\frac{3}{5} + \frac{4}{5} = \frac{7}{5}$$

2. **Switch improper fractions to mixed numbers, write down the fractional part, and carry over the whole number.**

   Because the sum is an improper fraction, convert it to the mixed number 1⅖ (flip to Chapter 9 for more on converting improper fractions to mixed numbers). Write down ⅖ and carry the 1 over to the whole-number column.

3. **Add the whole number parts, including any whole numbers you carried over when you switched to a mixed number.**

$$1 + 8 + 6 = 15$$

Here's how the solved problem looks in column form. (Be sure to line up the whole numbers in one column and the fractions in another.)

$$
\begin{array}{r}
1 \\
8\,{}^{3}\!/_{5} \\
+\ 6\,{}^{4}\!/_{5} \\
\hline
15\,{}^{2}\!/_{5}
\end{array}
$$

As with any other problems involving fractions, sometimes you need to reduce at the end of Step 1.

The same basic idea works no matter how many mixed numbers you want to add. For example, suppose you want to add 5⅘ + 11⅞ + 3⅘ + 1⅝:

1. **Add the fractions.**

$$\frac{4}{9} + \frac{7}{9} + \frac{8}{9} + \frac{5}{9} = \frac{24}{9}$$

2. **Switch improper fractions to mixed numbers, write down the fractional part, and carry over the whole number.**

   Because the result is an improper fraction, convert it to the mixed number 2⁶⁄₉ and then reduce it to 2⅔ (for more on converting and reducing fractions, see Chapter 9). I recommend doing these calculations on a piece of scrap paper.

   Write down ⅔ and carry the 2 to the whole-number column.

3. **Add the whole numbers.**

$$2 + 5 + 11 + 3 + 1 = 22$$

Here's how the problem looks after you solve it:

$$
\begin{array}{r}
2 \\
5\frac{1}{9} \\
11\frac{7}{9} \\
3\frac{8}{9} \\
+\ 1\frac{5}{9} \\
\hline
22\frac{2}{3}
\end{array}
$$

Summing up mixed numbers when the denominators are different

The most difficult type of mixed number addition is when the denominators of the fractions are different. This difference doesn't change Steps 2 or 3, but it does make Step 1 tougher.

For example, suppose you want to add $16\frac{3}{5}$ and $7\frac{7}{9}$.

1. **Add the fractions.**

   Add $\frac{3}{5}$ and $\frac{7}{9}$. You can use any method from earlier in this chapter. Here, I use the easy way:

   $$\frac{3}{5} + \frac{7}{9} = \frac{3 \cdot 9 + 7 \cdot 5}{5 \cdot 9} = \frac{27 + 35}{45} = \frac{62}{45}$$

2. **Switch improper fractions to mixed numbers, write down the fractional part, and carry over the whole number.**

   This fraction is improper, so change it to the mixed number $1\frac{17}{45}$. Fortunately, the fractional part of this mixed number isn't reducible.

   Write down the $\frac{17}{45}$ and carry over the 1 to the whole-number column.

3. **Add the whole numbers.**

   $$1 + 16 + 7 = 24$$

Here's how the completed problem looks:

$$
\begin{array}{r}
1 \\
16\frac{3}{5} \\
+\ 7\frac{7}{9} \\
\hline
24\frac{17}{45}
\end{array}
$$

Subtracting mixed numbers

The basic way to subtract mixed numbers is close to the way you add them. Again, the subtraction looks more like what you're used to with whole numbers. Here's how to subtract two mixed numbers:

1. **Find the difference of the fractional parts using any method you like.**

2. **Find the difference of the two whole number parts.**

Along the way, though, you may encounter a couple more twists and turns. I keep you on track so that by the end of this section, you can do any mixed-number subtraction problem.

Taking away mixed numbers when the denominators are the same

As with addition, subtraction is much easier when the denominators are the same. For example, suppose you want to subtract 7⅗ – 3⅕. Here's what the problem looks like in column form:

$$
\begin{array}{r}
7\,\tfrac{3}{5} \\
-\ 3\,\tfrac{1}{5} \\
\hline
4\,\tfrac{2}{5}
\end{array}
$$

In this problem, I subtract ⅗ – ⅕ = ⅖. Then I subtract 7 – 3 = 4. Not too terrible, agreed?

One complication arises when you try to subtract a larger fractional part from a smaller one. Suppose you want to find 11⅙ – 2⅚. This time, if you try to subtract the fractions, you get

$$\tfrac{1}{6} - \tfrac{5}{6} = -\tfrac{4}{6}$$

Obviously, you don't want to end up with a negative number in your answer. You can handle this problem by borrowing from the column to the left. This idea is very similar to the borrowing that you use in regular subtraction, with one key difference.

When borrowing in mixed-number subtraction,

1. **Borrow 1 from the whole-number portion and add it to the fractional portion, turning the fraction into a mixed number.**

   To find 11⅙ – 2⅚, borrow 1 from the 11 and add it to ⅙, making it the mixed number 1⅙:

   $$11\tfrac{1}{6} = 10 + 1\tfrac{1}{6}$$

2. **Change this new mixed number into an improper fraction.**

   Here's what you get when you change 1⅙ into an improper fraction:

   $$10 + 1\tfrac{1}{6} = 10\tfrac{7}{6}$$

   The result is 10⅞. This answer is a weird cross between a mixed number and an improper fraction, but it's what you need to handle the job.

3. **Use the result in your subtraction.**

   $$
   \begin{array}{r}
   10\,\tfrac{7}{6} \\
   -\ 2\,\tfrac{5}{6} \\
   \hline
   8\,\tfrac{2}{6}
   \end{array}
   $$

In this case, you have to reduce the fractional part of the answer:

$$8\tfrac{2}{6} = 8\tfrac{1}{3}$$

Subtracting mixed numbers when the denominators are different

Subtracting mixed numbers when the denominators are different is just about the hairiest thing you're ever going to have to do in pre-algebra. Fortunately, though, if you work through this chapter, you have all the skills you need.

Suppose you want to subtract $15\tfrac{4}{11} - 12\tfrac{3}{7}$. Because the denominators are different, subtracting the fractions becomes more difficult. But you have another question to think about: In this problem, do you need to borrow? If $\tfrac{4}{11}$ is greater than $\tfrac{3}{7}$, you don't have to borrow. But if $\tfrac{4}{11}$ is less than $\tfrac{3}{7}$, you do. (For more on borrowing in mixed number subtraction, see the preceding section.)

In Chapter 9, I show you how to test two fractions to see which is greater by cross-multiplying:

$$\frac{4}{11} \bowtie \frac{3}{7}$$
$$4 \cdot 7 = 28 \qquad 3 \cdot 11 = 33$$

Because 28 is less than 33, $\tfrac{4}{11}$ is less than $\tfrac{3}{7}$, so you do have to borrow. I get the borrowing out of the way first:

$$15\frac{4}{11} = 14 + 1\frac{4}{11} = 14\frac{15}{11}$$

Now the problem looks like this:

$$14\frac{15}{11} - 12\frac{3}{7}$$

The first step, subtracting the fractions, is going to be the most time-consuming, so as I show you earlier in "Subtracting fractions with different denominators," you can take care of that on the side:

$$\frac{15}{11} - \frac{3}{7} = \frac{15 \cdot 7 - 3 \cdot 11}{11 \cdot 7} = \frac{105 - 33}{77} = \frac{72}{77}$$

The good news is that this fraction can't be reduced. (They can't be reduced because 72 and 77 have no common factors: $72 = 2 \cdot 2 \cdot 2 \cdot 3 \cdot 3$, and $77 = 7 \cdot 11$.) So the hard part of the problem is done, and the rest follows easily:

$$
\begin{array}{r}
14\,^{15}\!/_{11} \\
-\ 12\,^{3}\!/_{7} \\
\hline
2\,^{72}\!/_{77}
\end{array}
$$

This problem is about as difficult as a mixed-number subtraction problem gets. Take a look over it step by step. Or better yet, copy the problem and then close the book and try to work through the steps on your own. If you get stuck, that's okay. Better now than on an exam!

Chapter 11

Dallying with Decimals

*B*ecause early humans used their fingers for counting, the number system is based on the number 10. That's why numbers come in ones, tens, hundreds, thousands, and so on. A *decimal* — with its handy decimal point — allows people to work with numbers smaller than one: tenths, hundredths, thousandths, and the like.

Here's some lovely news: Decimals are much easier to work with than fractions (which I discuss in Chapters 9 and 10). Decimals look and feel more like whole numbers than fractions do, so when working with decimals, you don't have to worry about reducing and increasing terms, improper fractions, mixed numbers, and a lot of other stuff.

Performing the Big Four operations — addition, subtraction, multiplication, and division — on decimals is very close to performing them on whole numbers (which I cover in Part II of the book). The numerals 0 through 9 work just like they usually do. As long as you get the decimal point in the right place, you're home free.

In this chapter, I show you all about working with decimals. I also show you how to convert fractions to decimals and decimals to fractions. Finally, I give you a peek into the strange world of repeating decimals.

Basic Decimal Stuff

The good news about decimals is that they look a lot more like whole numbers than fractions do. So, a lot of what you find out about whole numbers in Chapter 2 applies to decimals as well. In this section, I introduce you to decimals, starting with place value.

When you understand place value of decimals, a lot falls into place. Then I discuss trailing zeros, and what happens when you move the decimal point either to the left or to the right.

Counting dollars and decimals

You use decimals all the time when you count money. And a great way to begin thinking about decimals is as dollars and cents. For example, you know that $0.50 is half of a dollar (see Figure 11-1), so this information tells you:

$0.5 = \frac{1}{2}$

Figure 11-1:
One half
(0.5) of a
dollar bill.

Notice that in the decimal 0.5, I drop the zero at the end. This is a common practice with decimals.

You also know that $0.25 is a quarter — that is, one-fourth of a dollar (see Figure 11-2) — so:

$0.25 = \frac{1}{4}$

Figure 11-2:
One-fourth
(0.25) of a
dollar bill.

Similarly, you know that $0.75 is three quarters, or three-fourths of a dollar (see Figure 11-3), so:

$0.75 = \frac{3}{4}$

Figure 11-3:
Three-
fourths
(0.75) of a
dollar bill.

Taking this idea even further, you can use the remaining denominations of coins — dimes, nickels, and pennies — to make further connections between decimals and fractions.

A dime = $0.10 = $\frac{1}{10}$ of a dollar, so $\frac{1}{10}$ = 0.1

A nickel = $0.05 = $\frac{1}{20}$ of a dollar, so $\frac{1}{20}$ = 0.05

A penny = $0.01 = $\frac{1}{100}$ of a dollar, so $\frac{1}{100}$ = 0.01

Notice that I again drop the zero in the decimal 0.1, but that I keep the zeros in the decimals 0.05 and 0.01. You can drop zeros from the end of a decimal, but you can't drop zeros that fall between the decimal point and another digit.

Decimals are just as good for cutting up cake as for cutting up money. Figure 11-4 gives you a look at the four cut-up cakes that I show you in Chapter 9. This time, I give you the decimals that tell you how much cake you have. Fractions and decimals accomplish the same task: allowing you to cut a whole object into pieces and talk about how much you have.

Figure 11-4:
Cakes cut and shaded into .75 (A), .4 (B), .1 (C), and .7 (D).

Place value of decimals

In Chapter 2, you find out about the place value of whole numbers. For example, Table 11-1 shows how the whole number 4,672 breaks down in terms of place value.

Table 11-1	Breaking Down 4,672 in Terms of Place Value		
Thousands	*Hundreds*	*Tens*	*Ones*
4	6	7	2

This number means 4,000 + 600 + 70 + 2.

With decimals, this idea is extended. First, a decimal point is placed to the right of the ones place in a whole number. Then, more numbers are appended to the right of the decimal point.

For example, the decimal 4,672.389 breaks down as shown in Table 11-2.

Table 11-2			Breaking Down the Decimal 4,672.389				
Thousands	*Hundreds*	*Tens*	*Ones*	*Decimal Point*	*Tenths*	*Hundredths*	*Thousandths*
4	6	7	2	.	3	8	9

This decimal means $4,000 + 600 + 70 + 2 + \frac{3}{10} + \frac{8}{100} + \frac{9}{1,000}$.

The connection between fractions and decimals becomes obvious when you look at place value. Decimals really are a shorthand notation for fractions. You can represent any fraction as a decimal.

Knowing the decimal facts of life

After you understand how place value works in decimals (as I explain in the previous section), a whole lot of facts about decimals begin to make sense. Two key ideas are trailing zeros and what happens when you move a decimal point left or right.

Trailing zeros

You probably know that you can attach zeros to the beginning of a whole number without changing its value. For example, these three numbers are all equal in value:

    27   027   0,000,027

The reason for this becomes clear when you know about place value of whole numbers. See Table 11-3.

Table 11-3			Example of Attaching Leading Zeros			
Millions	*Hundred Thousands*	*Ten Thousands*	*Thousands*	*Hundreds*	*Tens*	*Ones*
0	0	0	0	0	2	7

As you can see, 0,000,027 simply means 0 + 0 + 0 + 0 + 0 + 20 + 7. No matter how many zeros you add on to the beginning of a number, the number 27 doesn't change.

Zeros attached to the beginning of a number in this way are called *leading zeros*.

In decimals, this idea of zeros that don't add value to a number can be extended to trailing zeros.

A *trailing zero* is any zero that appears to the right of both the decimal point and every digit other than zero.

For example:

34.8   34.80   34.8000

All three of these numbers are the same. The reason becomes clear when you understand how place value works in decimals. See Table 11-4.

Table 11-4			Example of Attaching Trailing Zeros			
Tens	Ones	Decimal Point	Tenths	Hundredths	Thousandths	Ten Thousandths
3	4	.	8	0	0	0

In this example, 34.8000 means $30 + 4 + \frac{8}{10} + 0 + 0 + 0$

You can attach or remove as many trailing zeros as you want to without changing the value of a number.

When you understand trailing zeros, you can see that every whole number can be changed to a decimal easily. Just attach a decimal point and a 0 to the end of it. For example:

4 = 4.0

20 = 20.0

971 = 971.0

Make sure that you don't attach or remove any nonleading or nontrailing zeros, because doing this changes the value of the decimal.

For example, look at this number:

0450.0070

In this number, you can remove the leading and trailing zeros without changing the value, as follows:

450.007

The remaining zeros, however, need to stay where they are as *placeholders* between the decimal point and digits other than zero. See Table 11-5.

Table 11-5				Example of Zeros as Placeholders				
Thou-sands	Hun-dreds	Tens	Ones	Decimal Point	Tenths	Hun-dredths	Thou-sandths	Ten Thou-sandths
0	4	5	0	.	0	0	7	0

I continue to discuss zeros as placeholders in the next section.

Moving the decimal point

When you're working with whole numbers, you can multiply any number by 10 just by adding a zero to the end of it. For example:

45,971 · 10 = 459,710

To see why this answer is so, again think about the place value of digits and look at Table 11-6.

Table 11-6			Example, Decimal Points and Place Value of Digits			
Millions	Hundred Thousands	Ten Thousands	Thousands	Hundreds	Tens	Ones
		4	5	9	7	1
	4	5	9	7	1	0

Here's what these two numbers really mean:

45,971 = 40,000 + 5,000 + 900 + 70 + 1

459,710 = 400,000 + 50,000 + 9,000 + 700 + 10 + 0

As you can see, that little zero makes a big difference, because it causes the rest of the numbers to shift over one place.

This concept makes even more sense when you think about the decimal point. See Table 11-7.

Table 11-7		Example, Numbers Shifting Over One Place						
Hundred Thousands	Ten Thousands	Thousands	Hundreds	Tens	Ones	Decimal Point	Tenths	Hundredths
	4	5	9	7	1	.	0	0
4	5	9	7	1	0	.	0	0

In effect, adding a 0 to the end of a whole number moves the decimal point one place to the right. So, for any decimal, when you move the decimal point one place to the right, you multiply that number by 10. This becomes clear when you start with a simple number like 7:

7.0

70.0

700.0

7,000.0

In this case, the net effect is that you moved the decimal point three places to the right, which is the same as multiplying 7 by 1,000.

Similarly, to divide any number by 10, move the decimal point one place to the left. For example:

7.0

0.7

0.07

0.007

This time, the net effect is that you moved the decimal point three places to the left, which is the same as dividing 7 by 1,000.

Rounding decimals

Rounding decimals works almost exactly the same as rounding numbers. You'll use this skill when dividing decimals later in the chapter. Most commonly, you need to round a decimal either to a whole number or to one or two decimal places.

To round a decimal to a whole number, focus on the ones digit and the tenths digit. Round the decimal either up or down to the *nearest* whole number, dropping the decimal point:

    7.1 -> 7   32.9 -> 33   184.3 -> 184

When the tenths digit is 5, round the decimal *up*:

    83.5 -> 84   296.5 -> 297   1,788.5 -> 1,789

If the decimal has other decimal digits, just drop them:

    18.47 -> 18   21.618 -> 22   3.1415927 -> 3

Occasionally, a small change to the ones digits affects the other digits. (This may remind you of when the odometer in your car rolls a bunch of 9s over to 0s):

    99.9 -> 100   999.5 -> 1,000   99,999.712 -> 100,000

The same basic idea applies to rounding a decimal to any number of places. For example, to round a decimal to one decimal place, focus on the first and second decimal places (that is, the tenths and hundredths places):

    76.543 -> 76.5   100.6822 -> 100.7   10.10101 -> 10.1

To round a decimal to two decimal places, focus on the second and third decimal places (that is, the hundredths and thousandths places):

    444.4444 -> 444.44   26.55555 -> 26.56   99.997 -> 100.00

Performing the Big Four with Decimals

Everything you already know about adding, subtracting, multiplying, and dividing whole numbers (see Chapter 3) carries over when you work with decimals. In fact, in each case, there's really only one key difference: how to handle that pesky little decimal point. In this section, I show you how to perform the Big Four math operations with decimals.

The most common use of adding and subtracting decimals is when you're working with money — for example, balancing your checkbook. Later in this book, you find that multiplying and dividing by decimals is useful for calculating percentages (see Chapter 12), using scientific notation (see Chapter 14), and measuring with the metric system (see Chapter 15).

Adding decimals

Adding decimals is almost as easy as adding whole numbers. As long as you set up the problem correctly, you're in good shape. To add decimals, follow these steps:

1. **Line up the decimal points.**

2. **Add as usual from right to left, column by column.**

3. **Place the decimal point in the answer in line with the other decimal points in the problem.**

For example, suppose you want to add the numbers 14.5 and 1.89. Line up the decimal points neatly as follows:

$$14.5$$
$$\underline{+\ 1.89}$$

Begin adding from the right-hand column. Treat the blank space after 14.5 as a 0 — you can write this in as a trailing 0 (see earlier in this chapter to see why adding zeros to the end of a decimal doesn't change its value). Adding this column gives you 0 + 9 = 9:

$$14.50$$
$$\underline{+\ 1.89}$$
$$9$$

Continuing to the left, 5 + 8 = 13, so put down the 3 and carry the 1:

$$\overset{1}{14.50}$$
$$\underline{+\ 1.89}$$
$$39$$

Complete the problem column by column, and at the end, put the decimal point directly below the others in the problem:

$$
\begin{array}{r}
14.50 \\
+\ 1.89 \\
\hline
16.39
\end{array}
$$

When adding more than one decimal, the same rules apply. For example, suppose you want to add 15.1 + 0.005 + 800 + 1.2345. The most important idea is lining up the decimal points correctly:

$$
\begin{array}{r}
15.1 \\
0.005 \\
800.0 \\
+\ 1.2345 \\
\hline
\end{array}
$$

To avoid mistakes, be especially neat when adding a lot of decimals.

Because the number 800 isn't a decimal, I place a decimal point and a 0 at the end of it to be clear about how to line it up. If you like, you can make sure all numbers have the same number of decimal places (in this case, four) by adding trailing zeros. After you properly set up the problem, the addition is no more difficult than in any other addition problem:

$$
\begin{array}{r}
15.1000 \\
0.0050 \\
800.0000 \\
+\ 1.2345 \\
\hline
816.3395
\end{array}
$$

Subtracting decimals

Subtracting decimals uses the same trick as adding them (which I talk about in the preceding section). Here's how you subtract decimals:

1. **Line up the decimal points.**
2. **Subtract as usual from right to left, column by column.**
3. **When you're done, place the decimal point in the answer in line with the other decimal points in the problem.**

For example, suppose you want to figure out 144.87 − 0.321. First, line up the decimal points:

144.870

− 0.321

In this case, I add a zero at the end of the first decimal. This placeholder reminds you that in the right-hand column, you need to borrow to get the answer to 0 − 1:

$$
\begin{array}{r}
6\\
1\,4\,4.\,8\,\cancel{7}\,10\\
-\quad 0.\,3\,2\quad 1\\
\hline
4\quad 9
\end{array}
$$

The rest of the problem is very straightforward. Just finish the subtraction and drop the decimal point straight down:

$$
\begin{array}{r}
6\\
1\,4\,4.\,8\,\cancel{7}\,10\\
-\quad 0.\,3\,2\quad 1\\
\hline
1\,4\,4.\,5\,4\quad 9
\end{array}
$$

As with addition, the decimal point in the answer goes directly below where it appears in the problem.

Multiplying decimals

Multiplying decimals is different from adding and subtracting them in that you don't have to worry about lining up the decimal points (see the preceding sections). In fact, the only difference between multiplying whole numbers and decimals comes at the very end.

Here's how to multiply decimals:

1. **Perform the multiplication as you would for whole numbers.**

2. **When you're done, count the number of digits to the right of the decimal point in each factor and add the result.**

3. **Place the decimal point in your answer so that your answer has the same number of digits after the decimal point.**

This process sounds tricky, but multiplying decimals can actually be simpler than adding or subtracting them. Suppose, for instance, you want to multiply 23.5 by 0.16. The first step is to pretend that you're multiplying numbers without decimal points:

$$
\begin{array}{r}
23.5 \\
\times\,0.16 \\
\hline
1410 \\
2350 \\
\hline
3760
\end{array}
$$

This answer isn't complete, though, because you still need to find out where the decimal point goes. To do this, notice that 23.5 has one digit after the decimal point and that 0.16 has two digits after the decimal point. Because 1 + 2 = 3, place the decimal point in the answer so that it has three digits after the decimal point. (You can put your pencil at the 0 at the end of 3760 and move the decimal point three places to the left.)

$$
\begin{array}{rl}
23.5 & \text{1 digit after the decimal point} \\
\times\,0.16 & \text{2 digits after the decimal point} \\
\hline
1410 & \\
2350 & \\
\hline
3.760 & \text{1 + 2 = 3 digits after the decimal point}
\end{array}
$$

Even though the last digit in the answer is a 0, you still need to count this as a digit when placing the decimal point. After the decimal point is in place, you can drop trailing zeros (flip to "Basic Decimal Stuff" earlier in this chapter to see why the zeros at the end of a decimal that don't change the value of the number).

So the answer is 3.760, which is equal to 3.76.

Dividing decimals

Long division has never been a crowd pleaser. Dividing decimals is almost the same as dividing whole numbers, which is why lots of people don't particularly like dividing decimals, either.

But at least you can take comfort in the fact that when you know how to do long division (which I cover in Chapter 3), figuring out how to divide decimals is easy. The main difference comes at the beginning, before you start dividing.

Here's how to divide decimals:

1. **Turn the _divisor_ (the number you're dividing by) into a whole number by moving the decimal point all the way to the right; at the same time, move the decimal point in the _dividend_ (the number you're dividing) the same number of places to the right.**

   For example, suppose you want to divide 10.274 by 0.11. Write the problem as usual:

   $$0.11\overline{)10.274}$$

   Turn 0.11 into a whole number by moving the decimal point in 0.11 two places to the right, giving you 11. At the same time, move the decimal point in 10.274 two places to the right, giving you 1,027.4:

   $$11.\overline{)1027.4}$$

2. **Place a decimal point in the _quotient_ (the answer) directly above where the decimal point now appears in the dividend.**

   Here's what this step looks like:

   $$11.\overline{)1027.\overset{.}{4}}$$

3. **Divide as usual, being careful to line up the quotient properly so that the decimal point falls into place.**

   To start out, notice that 11 is too large to go into either 1 or 10. However, 11 does go into 102 (9 times). So write the first digit of the quotient just above the 2 and continue:

   $$
   \begin{array}{r}
   9\,. \\
   11.\overline{)1027.4} \\
   \underline{99} \\
   37
   \end{array}
   $$

   I left off after bringing down the next number, 7. This time, 11 goes into 37 three times. The important thing is to place the next digit in the answer just above the 7:

   $$
   \begin{array}{r}
   93. \\
   11.\overline{)1027.4} \\
   \underline{99} \\
   37 \\
   \underline{33} \\
   44
   \end{array}
   $$

I left off after bringing down the next number, 4. Now, 11 goes into 44 four times. Again, be careful to place the next digit in the quotient just above the 4, and complete the division:

$$
\begin{array}{r}
93.4 \\
11. \overline{)1027.4} \\
99 \\
\hline
37 \\
33 \\
\hline
44 \\
44 \\
\hline
0
\end{array}
$$

So the answer is 93.4. As you can see, as long as you're careful when placing the decimal point and the digits, the correct answer appears with the decimal point in the right position.

Dealing with more zeros in the dividend

Sometimes, you may have to add one or more trailing zeros to the dividend. As I discuss earlier in this chapter, you can add as many trailing zeros as you like to a decimal without changing its value. For example, suppose you want to divide 67.8 by 0.333:

$$0.333\,\overline{)67.8}$$

1. **Change 0.333 into a whole number by moving the decimal point three places to the right; at the same time, move the decimal point in 67.8 three places to the right:**

$$333.\,\overline{)67800.}$$

   In this case, when you move the decimal point in 67.8, you run out of room, so you have to add a couple of zeros to the dividend. This step is perfectly valid, and you need to do this whenever the divisor has more decimal places than the dividend.

2. **Place the decimal point in the quotient directly above where it appears in the dividend:**

$$333.\,\overline{)67800.}$$

3. **Divide as usual, being careful to line up the numbers in the quotient correctly. This time, 333 doesn't go into 6 or 67, but it does go into 678 (two times). So place the first digit of the quotient directly above the 8:**

$$
\begin{array}{r}
2 \\
333.\,\overline{)67800.} \\
666 \\
\hline
120
\end{array}
$$

I've jumped forward in the division to the place where I bring down the first 0. At this point, 333 doesn't go into 120, so you need to put a 0 above the first 0 in 67,800 and bring down the second 0. Now, 333 does go into 1,200, so place the next digit in the answer (3) over the second 0:

$$
\begin{array}{r}
203. \\
333. \overline{)\,67800.} \\
666 \\
\overline{1200} \\
999 \\
\overline{201}
\end{array}
$$

This time, the division doesn't work out evenly. If this were a problem with whole numbers, you'd finish by writing down a remainder of 201. (For more on remainders in division, see Chapter 1.) But decimals are a different story. The next section explains why, with decimals, the show must go on.

Completing decimal division

When you're dividing whole numbers, you can complete the problem simply by writing down the remainder. But remainders are *never* allowed in decimal division.

A common way to complete a problem in decimal division is to round off the answer. In most cases, you'll be instructed to round your answer to the nearest whole number or to one or two decimal places (see earlier in this chapter to find out how to round off decimals.)

To complete a decimal division problem by rounding it off, you need to add at least one trailing zero to the dividend. To round a decimal to

✔ a whole number, add one trailing zero

✔ one decimal place, add two trailing zeros

✔ two decimal places, add three trailing zeros

Here's what the problem looks like with a trailing zero attached:

$$
\begin{array}{r}
203. \\
333. \overline{)\,67800.0} \\
666 \\
\overline{1200} \\
999 \\
\overline{2010}
\end{array}
$$

Attaching a trailing zero doesn't change a decimal, but it does allow you to bring down one more number, changing 201 into 2,010. Now you can divide 333 into 2,010:

$$
\begin{array}{r}
203.6 \\
333. \overline{)67800.0} \\
\underline{666} \\
1200 \\
\underline{999} \\
2010 \\
\underline{1998} \\
12
\end{array}
$$

At this point, the you can round the answer to the nearest whole number, 204. I give you more practice dividing decimals later in this chapter.

Converting between Decimals and Fractions

Fractions (see Chapters 9 and 10) and decimals are similar in that they both allow you to represent parts of the whole — that is, these numbers fall on the number line *between* whole numbers.

In practice, though, sometimes one of these options is more desirable than the other. For example, calculators love decimals, but aren't so crazy about fractions. To use your calculator, you may have to change fractions into decimals.

As another example, some units of measurement (such as inches) use fractions while others (such as meters) use decimals. To change units, you may need to convert between fractions and decimals.

In this section, I show you how to convert back and forth between fractions and decimals. (If you need a refresher on fractions, review chapters 9 and 10 before proceeding.)

Making simple conversions

Some decimals are so common that you should memorize how to represent them as fractions. Here's how to convert all the one-place decimals to fractions:

.1	.2	.3	.4	.5	.6	.7	.8	.9
1/10	1/5	3/10	2/5	1/2	3/5	7/10	4/5	9/10

And, here are few more common decimals that translate easily to fractions:

.125	.25	.375	.625	.75	.875
1/8	1/4	3/8	5/8	3/4	7/8

Changing decimals to fractions

Converting a decimal to a fraction is pretty simple. The only tricky part comes in when you have to reduce the fraction or change it to a mixed number.

In this section, I first show you the easy case, when no further work is necessary. Then I show you the harder case, when you need to tweak the fraction. I also show you a great time-saving trick.

Doing a basic decimal-to-fraction conversion

Here's how to convert a decimal to a fraction:

1. **Draw a line (fraction bar) under the decimal and place a 1 underneath it.**

   Suppose you want to turn the decimal 0.3763 into a fraction. Draw a line under 0.3763 and place a 1 underneath it:

   $$\frac{0.3763}{1}$$

   This number looks like a fraction, but technically it isn't one because the top number (the numerator) is a decimal.

2. **Move the decimal point one place to the right and add a 0 after the 1.**

   $$\frac{3.763}{10}$$

3. **Repeat Step 2 until the decimal point moves all the way to the right, so you can drop the decimal point entirely.**

   In this case, this is a three-step process:

   $$\frac{37.63}{100} = \frac{376.3}{1,000} = \frac{3,763}{10,000}$$

As you can see on the last step, the decimal point in the numerator moves all the way to the end of the number, so dropping the decimal point is okay.

Note: Moving a decimal point one place to the right is the same thing as multiplying a number by 10. When you move the decimal point four places in this problem, you're essentially multiplying the 0.3763 and the 1 by 10,000. Notice that the number of digits after the decimal point in the original decimal is equal to the number of 0s that end up following the 1.

4. **If necessary, change the resulting fraction to a mixed number and/or reduce it.**

   The fraction $\frac{3,763}{10,000}$ may seem like a large number, but it's smaller than 1 because the numerator is less than the denominator (bottom number). So this fraction is a proper fraction and can't be changed into a mixed number (see Chapter 10 to find out more about mixed numbers). This fraction is also in the lowest possible terms (you can't reduce it — see Chapter 9), so this problem is complete.

In the following section, I show you how to convert decimals to fractions when you have to work with mixed numbers and reduce the terms.

On better terms: Mixing numbers and reducing fractions

In some cases, you may have to reduce a fraction to lowest terms after you convert it (see Chapter 9 for more on reducing fractions). Reducing in these cases is usually easy, because no matter how large the denominator is, it's a multiple of 10, which has only two prime factors: 2 and 5 (see Chapter 8 for more on prime factors).

When you're converting a decimal to a fraction, if the decimal ends in an even number or in 5, you can reduce the fraction; otherwise, you can't. Also, if you have a number greater than 0 somewhere before the decimal point, you'll end up with a mixed number.

Suppose you want to convert the decimal 12.16 to a fraction. Do the following:

1. **Draw a line (fraction bar) under 12.16 and place a 1 underneath it:**

   $$\frac{12.16}{1}$$

2. **Move the decimal point one place to the right and add a 0 after the 1:**

   $$\frac{121.6}{10}$$

3. **Repeat Step 2:**

   $$\frac{1,216}{100}$$

4. **If necessary, change $^{1,216}/_{100}$ to a mixed number and reduce it to lowest terms.**

   This time, the numerator is greater than the denominator, so

   $$\frac{1,216}{100}$$

   is an improper fraction and must be changed to a mixed number.

After you've changed a decimal to an improper fraction, you can convert it to a mixed number by "pulling out" the original whole number part of the decimal from the fraction.

In this case, the original decimal was 12.16, so its whole number part was 12. Doing the following is perfectly okay:

$$\frac{1,216}{100} = 12\frac{16}{100}$$

This trick works only when you've changed a decimal to a fraction — don't try it with other fractions, or you'll get a wrong answer.

The mixed number $12^{16}/_{100}$ can be also be reduced as follows (if this last step seems unclear, check out Chapter 9 for more on reducing fractions):

$$12\frac{16}{100} = 12\frac{8}{50} = 12\frac{4}{25}$$

Changing fractions to decimals

Converting fractions to decimals isn't difficult, but to do it, you need to know about decimal division. If you need to get up to speed on this, check out "Dividing decimals" earlier in this chapter.

To convert a fraction to a decimal:

1. Set up the fraction as a decimal division, dividing the numerator (top number) by the denominator (bottom number).

2. Attach enough trailing zeros to the numerator so that you can continue dividing until you find that the answer is either a *terminating decimal* or a *repeating decimal*.

Don't worry, I explain terminating and repeating decimals later.

The last stop: Terminating decimals

Sometimes, when you divide the numerator of a fraction by the denominator, the division eventually works out evenly. The result is a *terminating decimal*.

For example, suppose you want to change the fraction ⅖ to a decimal. Here's your first step:

$$5\overline{)2}$$

One glance at this problem, and it looks like you're doomed from the start because 5 doesn't go into 2. But watch what happens when I add a few trailing zeros. Notice that I also place another decimal point in the answer just above the first decimal point. This step is important — you can read more about it in "Dividing decimals":

$$5\overline{)2.000}$$

Now you can divide because although 5 doesn't go into 2, 5 does go into 20 four times:

$$
\begin{array}{r}
0.4 \\
5\overline{)2.000} \\
\underline{20} \\
0
\end{array}
$$

You're done! As it turns out, you only needed one trailing zero, so you can ignore the rest:

$$\frac{2}{5} = 0.4$$

Because the division worked out evenly, the answer is an example of a *terminating decimal*.

As another example, suppose you want to find out how to represent ⁷⁄₁₆ as a decimal. As earlier, I attach three trailing zeros:

$$
\begin{array}{r}
0.437 \\
16\overline{)7.000} \\
\underline{64} \\
60 \\
\underline{48} \\
120 \\
\underline{112} \\
8
\end{array}
$$

This time, three trailing zeros aren't enough to get my answer, so I'll attach a few more and continue:

$$
\begin{array}{r}
0.4375 \\
16\overline{)7.000000} \\
\underline{64} \\
60 \\
\underline{48} \\
120 \\
\underline{112} \\
80 \\
\underline{80} \\
0
\end{array}
$$

At last, the division works out evenly, so again the answer is a terminating decimal. Therefore, $\frac{7}{16} = 0.4375$.

The endless ride: Repeating decimals

Sometimes when you try to convert a fraction to a decimal, the division *never* works out evenly. The result is a *repeating decimal* — that is, a decimal that cycles through the same number pattern forever.

You may recognize these pesky little critters from your calculator, when an apparently simple division problem produces a long string of numbers.

For example, to change $\frac{2}{3}$ to a decimal, begin by dividing 2 by 3. As in the last section, start out by adding three trailing zeros and see where it leads:

$$
\begin{array}{r}
0.666 \\
3\overline{)2.000} \\
\underline{18} \\
20 \\
\underline{18} \\
20 \\
\underline{18} \\
2
\end{array}
$$

At this point, you still haven't found an exact answer. But you may notice that a repeating pattern has developed in the division. No matter how many trailing zeros you attach to the number 2, the same pattern will continue forever. This answer, 0.666..., is an example of a repeating decimal. You can write $\frac{2}{3}$ as

$$2/3 = 0.\overline{6}$$

The bar over the 6 means that in this decimal, the number 6 repeats forever. You can represent many simple fractions as repeating decimals. In fact, *every* fraction can be represented either as a repeating decimal or as a terminating decimal — that is, as an ordinary decimal that ends.

Now suppose you want to find the decimal representation of ⁵⁄₁₁. Here's how this problem plays out:

$$
\begin{array}{r}
0.4545 \\
11\overline{)5.0000} \\
\underline{44} \\
60 \\
\underline{55} \\
50 \\
\underline{44} \\
60 \\
\underline{55} \\
5
\end{array}
$$

This time, the pattern repeats every other number — 4, then 5, then 4 again, and then 5 again, forever. Attaching more trailing zeros to the original decimal will only string out this pattern indefinitely. So you can write

$$5/11 = 0.\overline{45}$$

This time, the bar is over both the 4 and the 5, telling you that these two numbers alternate forever.

Repeating decimals are an oddity, but they aren't hard to work with. In fact, as soon as you can show that a decimal division is repeating, you've found your answer. Just remember to place the bar only over the numbers that keep on repeating.

Some decimals never end and never repeat. You can't write them as fractions, so mathematicians have agreed on some shorter ways of naming them so writing them out doesn't take, well, forever.

Chapter 12

Playing with Percents

*L*ike whole numbers and decimals, percents are a way to talk about parts of a whole. The word *percent* means "out of 100." So if you have 50% of something, you have 50 out of 100. If you have 25% of it, you have 25 out of 100. Of course, if you have 100% of anything, you have all of it.

In this chapter, I show you how to work with percents. Because percents resemble decimals, I first show you how to convert numbers back and forth between percents and decimals. No worries — this switch is very easy to do. Next, I show you how to convert back and forth between percents and fractions — also not too bad. When you understand how conversions work, I show you the three basic types of percent problems, plus a method that makes the problems simple.

Making Sense of Percents

The word *percent* literally means "for 100," but in practice, it means closer to "out of 100." For example, suppose a school has exactly 100 children — 50 girls and 50 boys. You can say that "50 out of 100" children are girls — or you can shorten it to simply "50 percent." Even shorter than that, you can use the symbol %, which means *percent*.

Saying that 50% of the students are girls is the same as saying that ½ of them are girls. Or if you prefer decimals, it's the same thing as saying that 0.5 of all the students are girls. This example shows you that percents, like fractions and decimals, are just another way of talking about parts of the whole. In this case, the whole is the total number of children in the school.

You don't literally have to have 100 of something to use a percent. You probably won't ever really cut a cake into 100 pieces, but that doesn't matter. The values are the same. Whether you're talking about cake, a dollar, or a group of children, 50% is still half, 25% is still one-quarter, and 75% is still three-quarters, and so on.

Any percentage smaller than 100% means less than the whole — the smaller the percentage, the less you have. You probably know this fact well from the school grading system. If you get 100%, you get a perfect score. And 90% is usually A work, 80% is a B, 70% is a C, and, well, you know the rest.

Of course, 0% means "0 out of 100" — any way you slice it, you have nothing.

Dealing with Percents Greater than 100%

100% means "100 out of 100" — in other words, everything. So when I say I have 100% confidence in you, I mean that I have complete confidence in you.

What about percentages more than 100%? Well, sometimes percentages like these don't make sense. For example, you can't spend more than 100% of your time playing basketball no matter how much you love the sport; 100% is all the time you have, and there ain't no more.

But lots of times, percentages larger than 100% are perfectly reasonable. For example, suppose I own a hot dog wagon and I sell the following:

> 10 hot dogs in the morning

> 30 hot dogs in the afternoon

The number of hot dogs I sell in the afternoon is 300% of the number I sold in the morning. That's three times as many.

Here's another way of looking at this: I sell 20 more hot dogs in the afternoon than in the morning, so this is a 200% *increase* in the afternoon — 20 is twice as many as 10.

Spend a little time thinking about this example until it makes sense. You visit some of these ideas again in Chapter 13, when I show you how to do word problems involving percents.

Converting to and from Percents, Decimals, and Fractions

To solve many percent problems, you need to change the percent to either a decimal or a fraction. Then you can apply what you know about solving decimal and fraction problems. That's why I show you how to convert to and from percents before I show you how to solve percent problems.

Percents and decimals are very similar ways of expressing parts of a whole. This similarity makes converting percents to decimals and vice versa mostly a matter of moving the decimal point. It's so simple you can probably do it in your sleep (but you should probably stay awake when you first read about the concept).

Percents and fractions both express the same idea — parts of a whole — in different ways. So converting back and forth between percents and fractions isn't quite as simple as just moving the decimal point back and forth. In this section, I cover the ways to convert to and from percents, decimals, and fractions, starting with percents to decimals.

Going from percents to decimals

To convert a percent to a decimal, drop the percent sign (%) and move the decimal point two places to the left. That's all there is to it. Remember that in a whole number, the decimal point comes at the end. For example,

$$2.5\% = 0.025$$
$$4\% = 0.04$$
$$36\% = 0.36$$
$$111\% = 1.11$$

Changing decimals into percents

To convert a decimal to a percent, move the decimal point two places to the right and add a percent sign (%):

$$0.07 = 7\%$$
$$0.21 = 21\%$$
$$0.375 = 37.5\%$$

Switching from percents to fractions

Converting percents to fractions is fairly straightforward. Remember that the word *percent* means "out of 100." So changing percents to fractions naturally involves the number 100.

To convert a percent to a fraction, use the number in the percent as your numerator (top number) and the number 100 as your denominator (bottom number):

$$39\% = \frac{39}{100} \qquad 86\% = \frac{86}{100} \qquad 217\% = \frac{217}{100}$$

As always with fractions, you may need to reduce to lowest terms or convert an improper fraction to a mixed number (flip to Chapter 9 for more on these topics).

In the three examples, $\frac{39}{100}$ can't be reduced or converted to a mixed number. However, $\frac{86}{100}$ can be reduced because the numerator and denominator are both even numbers:

$$\frac{86}{100} = \frac{43}{50}$$

And $\frac{217}{100}$ can be converted to a mixed number because the numerator (217) is greater than the denominator (100):

$$\frac{217}{100} = 2\frac{17}{100}$$

Once in a while, you may start out with a percentage that's a decimal such as 99.9%. The rule is still the same, but now you have a decimal in the numerator (top number), which most people don't like to see. To get rid of it, move the decimal point one place to the right in both the numerator and the denominator:

$$99.9\% = \frac{99.9}{100} = \frac{999}{1,000}$$

Thus, 99.9% converts to the fraction $\frac{999}{1,000}$.

Turning fractions into percents

Converting a fraction to a percent is really a two-step process. Here's how to convert a fraction to a percent:

1. **Convert the fraction to a decimal.**

   For example, suppose you want to convert the fraction ⅘ to a percent. To convert ⅘ to a decimal, you can divide the numerator by the denominator, as shown in Chapter 11:

   ⅘ = 0.8

2. **Convert this decimal to a percent.**

   Convert 0.8 to a percent by moving the decimal point two places to the right and adding a percent sign (as I show you earlier in "Changing decimals into percents").

   0.8 = 80%

Now suppose you want to convert the fraction ⅝ to a percent. Follow these steps:

1. **Convert ⅝ to a decimal by dividing the numerator by the denominator:**

$$
\begin{array}{r}
0.625 \\
8\overline{)5.000} \\
-48 \\
\hline
20 \\
-16 \\
\hline
40 \\
-40 \\
\hline
0
\end{array}
$$

   Therefore, ⅝ = 0.625.

2. **Convert 0.625 to a percent by moving the decimal point two places to the right and adding a percent sign (%):**

   0.625 = 62.5%

Solving Percent Problems

When you know the connection between percents and fractions, which I discuss earlier in "Converting to and from Percents, Decimals, and Fractions," you can solve a lot of percent problems with a few simple tricks. Others, however, require a bit more work. In this section, I show you how to tell an easy percent problem from a tough one, and I give you the tools to solve all of them.

Figuring out simple percent problems

A lot of percent problems turn out to be easy when you give them a little thought. In many cases, just remember the connection between percents and fractions and you're halfway home:

- **Finding 100% of a number:** Remember that 100% means the whole thing, so 100% of any number is simply the number itself:

    100% of 5 is 5

    100% of 91 is 91

    100% of 732 is 732

- **Finding 50% of a number:** Remember that 50% means half, so to find 50% of a number, just divide it by 2:

    50% of 20 is 10

    50% of 88 is 44

    50% of 7 is $\frac{7}{2}$ (or $3\frac{1}{2}$ or 3.5)

- **Finding 25% of a number:** Remember that 25% equals ¼, so to find 25% of a number, divide it by 4:

    25% of 40 is 10

    25% of 88 is 22

    25% of 15 is $\frac{15}{4}$ (or $3\frac{3}{4}$ or 3.75)

- **Finding 20% of a number:** Finding 20% of a number is handy if you like the service you've received in a restaurant, because a good tip is 20% of the check. Because 20% equals ⅕, you can find 20% of a number by dividing it by 5. But I can show you an easier way:

    To find 20% of a number, move the decimal point one place to the left and double the result:

    20% of 80 = 8 · 2 = 16

    20% of 300 = 30 · 2 = 60

    20% of 41 = 4.1 · 2 = 8.2

- **Finding 10% of a number:** Finding 10% of any number is the same as finding ¹⁄₁₀ of that number. To do this, just move the decimal point one place to the left:

    10% of 30 is 3

    10% of 41 is 4.1

    10% of 7 is 0.7

✔ **Finding 200%, 300%, and so on of a number:** Working with percents that are multiples of 100 is easy. Just drop the two 0s and multiply by the number that's left:

$$200\% \text{ of } 7 = 2 \cdot 7 = 14$$

$$300\% \text{ of } 10 = 3 \cdot 10 = 30$$

$$1,000\% \text{ of } 45 = 10 \cdot 45 = 450$$

(See the earlier "Dealing with Percents Greater than 100%" section for details on what having more than 100% really means.)

Turning the problem around

Here's a trick that makes certain tough-looking percent problems so easy that you can do them in your head. Simply move the percent sign from one number to the other and flip the order of the numbers.

Suppose someone wants you to figure out the following:

88% of 50

Finding 88% of anything isn't an activity that anybody looks forward to. But an easy way of solving the problem is to switch it around:

88% of 50 = 50% of 88

This move is perfectly valid, and it makes the problem a lot easier. As I discuss in the preceding section, "Figuring out simple percent problems," 50% of 88 is simply half of 88:

88% of 50 = 50% of 88 = 44

As another example, suppose you want to find

7% of 200

Again, finding 7% is tricky, but finding 200% is simple, so switch the problem around:

7% of 200 = 200% of 7

In the preceding section, I tell you that to find 200% of any number, you just multiply that number by 2:

$$7\% \text{ of } 200 = 200\% \text{ of } 7 = 2 \cdot 7 = 14$$

Deciphering more-difficult percent problems

You can solve a lot of percent problems using the tricks I show you earlier in this chapter. But what about this problem?

35% of 80 = ?

Ouch — this time, the numbers you're working with aren't so friendly. When the numbers in a percent problem become a little more difficult, the tricks no longer work, so you want to know how to solve *all* percent problems.

Here's how to find any percent of any number:

1. **Change the word *of* to a multiplication sign and the percent to a decimal (as I show you earlier in this chapter).**

   Changing the word *of* to a multiplication sign is a simple example of turning words into numbers, as I discuss in Chapters 6 and 13. This change turns something unfamiliar into a form that you know how to work with.

   Suppose you want to find 35% of 80. Here's how you start:

   35% of 80 = 0.35 · 80

2. **Solve the problem using decimal multiplication (see Chapter 11).**

   Here's what the example looks like:

$$\begin{array}{r} 0.35 \\ \times\ \ 80 \\ \hline 28.00 \end{array}$$

   So 35% of 80 is 28.

As another example, suppose you want to find 12% of 31. Again, start by changing the percent to a decimal and the word *of* to a multiplication sign:

12% of 31 = 0.12 · 31

Now you can solve the problem with decimal multiplication:

$$\begin{array}{r} 0.12 \\ \times\ \ 31 \\ \hline 12 \\ 360 \\ \hline 3.72 \end{array}$$

So 12% of 31 is 3.72.

Putting All the Percent Problems Together

In the preceding section, "Solving Percent Problems," I give you a few ways to find any percent of any number. This type of percent problem is the most common — that's why it gets top billing.

But percents are commonly used in a wide range of business applications such as banking, real estate, payroll, and taxes. (I show you some real-world applications when I discuss word problems in Chapter 13.) And depending on the situation, two other common types of percent problems may present themselves.

In this section, I show you these two additional types of percent problems and how they relate to the type you now know how to solve. I also give you a simple tool to make quick work of all three types.

Identifying the three types of percent problems

Earlier in this chapter, I show you how to solve problems that look like this:

50% of 2 is ?

And the answer, of course, is 1. (See "Solving Percent Problems" for details on how to get this answer.) Given two pieces of information — the percent and the number to start with — you can figure out what number you end up with.

Now suppose instead that I leave out the percent but give you the starting and ending numbers:

? % of 2 is 1

You can still fill in the blank without too much trouble. Similarly, suppose that I leave out the starting number but give the percent and the ending number:

50% of ? is 1

Again, you can fill in the blank.

If you get this basic idea, you're ready to solve percent problems. When you boil them down, nearly all percent problems are like one of the three types I show in Table 12-1.

Table 12-1	The Three Main Types of Percent Problems	
Problem Type	*What to Find*	*Example*
Type #1	The ending number	50% of 2 is <u>what</u>?
Type #2	The percentage	<u>What</u> percent of 2 is 1?
Type #3	The starting number	50% of <u>what</u> is 1?

In each case, the problem gives you two of the three pieces of information, and your job is to figure out the remaining piece. In the next section, I give you a simple tool to help you solve all three of these types of percent problems.

Introducing the percent circle

The *percent circle* is a simple visual aid that helps you make sense of percent problems so that you can solve them easily. The trick to using a percent circle is to write information into it. For example, Figure 12-1 shows how to record the information that 50% of 2 is 1.

Figure 12-1:
In the per-
cent circle,
the ending
number is at
the top, the
percentage
is at the
left, and the
starting
number is at
the right.

Notice that as I fill in the percent circle, I change the percentage, 50%, to its decimal equivalent, 0.5 (for more on changing percents to decimals, see "Going from percents to decimals" earlier in this chapter).

Here are the two main features of the percent circle:

- ✔ When you multiply the two bottom numbers together, they equal the top number:

    $$0.5 \cdot 2 = 1$$

- ✔ If you make a fraction out of the top number and either bottom number, that fraction equals the *other* bottom number:

    $$\frac{1}{2} = 0.5 \quad \text{and} \quad \frac{1}{0.5} = 2$$

These features are the heart and soul of the percent circle. They enable you to solve any of the three types of percent problems quickly and easily.

Most percent problems give you enough information to fill in two of the three sections of the percent circle. But no matter which two sections you fill in, you can find out the number in the third section.

Finding the ending number

Suppose you want to find out the answer to this problem:

> What is 75% of 20?

You're given the percent and the starting number and asked to find the ending number. To use the percent circle on this problem, fill in the information as shown in Figure 12-2.

Figure 12-2: Putting 75% of 20 into a percent circle

Because 0.75 and 20 are both bottom numbers in the circle, multiply them to get the answer:

$$
\begin{array}{r}
0.75 \\
\times\ 20 \\
\hline
15.00
\end{array}
$$

So 75% of 20 is 15.

As you can see, this method is essentially the same one I show you earlier in this chapter in "Deciphering more difficult percent problems," where you translate the word *of* as a multiplication sign. You still use multiplication to get your answer, but with the percent circle, you're less likely to get confused.

Discovering the percentage

In the second type of problem, I give you both the starting and ending numbers, and I ask you to find the percentage. Here's an example:

What percent of 50 is 35?

In this case, the starting number is 50 and the ending number is 35. Set up the problem on the percent circle as shown in Figure 12-3.

Figure 12-3:
Determining what percent of 50 is 35.

This time, 35 is above 50, so make a fraction out of these two numbers:

$$\frac{35}{50}$$

This fraction *is your answer,* and all you have to do is convert the fraction to a percent as I discuss earlier in this chapter in "Turning fractions into percents." First, convert $\frac{35}{50}$ to a decimal:

$$\begin{array}{r} 0.7 \\ 50\overline{)35.0} \\ \underline{350} \\ 0 \end{array}$$

Now convert 0.7 to a percent:

0.7 = 70%

Tracking down the starting number

In the third type of problem, you get the percentage and the ending number, and you have to find the starting number. For example,

15% of what number is 18?

This time, the percentage is 15% and the ending number is 18, so fill in the percent circle as shown in Figure 12-4.

Because 18 is above 0.15 in the circle, make a fraction out of these two numbers:

$$\frac{18}{0.15}$$

This fraction *is your answer;* you just need to convert to a decimal as I show you in Chapter 11:

$$
\begin{array}{r}
120. \\
(.)15.\,\overline{)1800.} \\
\underline{15} \\
30 \\
\underline{30} \\
0
\end{array}
$$

In this case, the "decimal" you find is the whole number 120, so 15% of 120 is 18.

Chapter 13

Word Problems with Fractions, Decimals, and Percents

*I*n Chapter 6, I show you how to solve word problems (also known as story problems) by setting up word equations that use the Big Four operations (adding, subtracting, multiplying, and dividing). In this chapter, I show you how to extend these skills to solve word problems with fractions, decimals, and percents.

First, I show you how to solve relatively easy problems, in which all you need to do is add or subtract fractions, decimals, or percents. Next, I show you how to solve problems that require you to multiply fractions. Such problems are easy to spot because they almost always contain the word *of*. After that, you discover how to solve percent problems by setting up a word equation and changing the percent to a decimal. Finally, I show you how to handle problems of percent increase and decrease. These problems are often practical money problems in which you figure out information about raises and salaries, costs and discounts, or amounts before and after taxes.

Adding and Subtracting Parts of the Whole in Word Problems

Certain word problems involving fractions, decimals, and percents are really just problems in adding and subtracting. You may add fractions, decimals, or percents in a variety of real-world settings that rely on weights and measures — such as cooking and carpentry. (In Chapter 15, I discuss these applications in depth.)

To solve these problems, you can use the skills that you pick up in Chapters 10 (for adding and subtracting fractions), 11 (for adding and subtracting decimals), and 12 (for adding and subtracting percents).

Sharing a pizza: Fractions

You may have to add or subtract fractions in problems that involve splitting up part of a whole. For example, consider the following:

> Joan ate ⅙ of a pizza, Tony ate ¼, and Sylvia ate ⅓. What fraction of the pizza was left when they were finished?

In this problem, just jot down the information that's given as word equations:

$$\text{Joan} = \frac{1}{6} \qquad \text{Tony} = \frac{1}{4} \qquad \text{Sylvia} = \frac{1}{3}$$

These fractions are part of one total pizza. To solve the problem, you need to find out how much all three people ate, so form the following word equation:

All three = Joan + Tony + Sylvia

Now you can substitute as follows:

$$\text{All three} = \frac{1}{6} + \frac{1}{4} + \frac{1}{3}$$

Chapter 10 gives you several ways to add these fractions. Here's one way:

$$\text{All three} = \frac{2}{12} + \frac{3}{12} + \frac{4}{12} = \frac{9}{12} = \frac{3}{4}$$

However, the question asks what fraction of the pizza was *left* after they finished, so you have to subtract that amount from the whole:

$$1 - \frac{3}{4} = \frac{1}{4}$$

Thus, the three people left ¼ of a pizza.

Buying by the pound: Decimals

You frequently work with decimals when dealing with money, metric measurements (see Chapter 15), and food sold by the pound. The following problem requires you to add and subtract decimals, which I discuss in Chapter 11. Even though the decimals may look intimidating, this problem is fairly simple to set up:

> Antonia bought 4.53 pounds of beef and 3.1 pounds of lamb. Lance bought 5.24 pounds of chicken and 0.7 pounds of pork. Which of them bought more meat, and how much more?

To solve this problem, you first find out how much each person bought:

> Antonia = 4.53 + 3.1 = 7.63
>
> Lance = 5.24 + 0.7 = 5.94

You can already see that Antonia bought more than Lance. To find how much more, subtract:

> 7.63 – 5.94 = 1.69

So Antonia bought 1.69 pounds more than Lance.

Splitting the vote: Percents

When percents represent answers in polls, votes in an election, or portions of a budget, the total often has to add up to 100%. In real life, you may see such info organized as a pie chart (which I discuss in Chapter 17). Solving problems about this kind of information often involves nothing more than adding and subtracting percents. Here's an example:

> In a recent mayoral election, five candidates were on the ballot. Faber won 39% of the vote, Gustafson won 31%, Ivanovich won 18%, Dixon won 7%, Obermayer won 3%, and the remaining votes went to write-in candidates. What percentage of voters wrote in their selection?

The candidates were in a single election, so all the votes have to total 100%. The first step here is just to add up the five percentages. Then subtract that value from 100%:

> 39% + 31% + 18% + 7% + 3% = 98%
>
> 100% – 98% = 2%

Because 98% of voters voted for one of the five candidates, the remaining 2% wrote in their selections.

Problems about Multiplying Fractions

In word problems, the word *of* almost always means *multiplication*. So whenever you see the word *of* following a fraction, decimal, or percent, you can usually replace it with a times sign.

When you think about it, *of* means multiplication even when you're not talking about fractions. For example, when you point to an item in a store and say, "I'll take three *of* those," in a sense you're saying, "I'll take that one *multiplied by* three."

The following examples give you practice turning word problems that include the word *of* into multiplication problems that you can solve with fraction multiplication.

When you divide up a single thing — such as one pizza or one death-by-chocolate cake — the word *of* still means to multiply; you're technically multiplying each fraction by 1. For example, the fraction that represents half of a pizza — that is, ½ of 1 pizza — is ½ · 1 = ½. Because anything times 1 is itself, you don't have to write the 1 at all — you can just add the fractions, as I do earlier in "Sharing a pizza: Fractions."

Renegade grocery shopping: Buying less than they tell you to

After you understand that the word *of* means multiplication, you have a powerful tool for solving word problems. For instance, you can figure out how much you'll spend if you don't buy food in the quantities listed on the signs. Here's an example:

> If beef costs $4 a pound, how much does ⅝ of a pound cost?

Here's what you get if you simply change the *of* to a multiplication sign:

$$\frac{5}{8} \cdot 1 \text{ pound of beef}$$

That's how much beef you're buying. However, you want to know the cost. Because the problem tells you that 1 pound = $4, you can replace *1 pound of beef* with *$4:*

$$= \frac{5}{8} \cdot \$4$$

Now you have an expression you can evaluate. Use the rules of multiplying fractions from Chapter 10 and solve:

$$= \frac{5 \cdot \$4}{8 \cdot 1} = \$\frac{20}{8}$$

This fraction reduces to $\$\frac{5}{2}$. However, the answer looks weird because dollars are usually expressed in decimals, not fractions. So convert this fraction to a decimal using the rules I show you in Chapter 11:

$$\$\frac{5}{2} = \$5 \div 2 = \$2.5$$

At this point, recognize that $2.5 is more commonly written as $2.50, and you have your answer.

Easy as pie: Working out what's left on your plate

Sometimes, when you're sharing something such as a pie, not everyone gets to it at the same time. The eager pie-lovers snatch the first slice, not bothering to divide the pie into equal servings, and the people who were slower or more patient or just not that hungry cut their own portions from what's left over. When someone takes a part of the leftovers, you can do a bit of multiplication to see how much of the *whole* pie that portion represents.

Consider the following example:

Jerry bought a pie and ate $\frac{1}{5}$ of it. Then his wife Doreen ate $\frac{1}{6}$ of what was left. How much of the total pie was left?

To solve this problem, begin by jotting down what the first sentence tells you:

$$\text{Jerry} = \frac{1}{5}$$

Doreen ate part of what was left, so write a word equation that tells you how much of the pie was left after Jerry was finished. He started with a whole pie, so subtract his portion from 1:

$$\text{Pie left after Jerry} = 1 - \frac{1}{5} = \frac{4}{5}$$

Next, Doreen ate $\frac{1}{6}$ of this amount. Rewrite the word *of* as multiplication and solve as follows. This answer tells you how much of the *whole* pie Doreen ate:

$$\text{Doreen} = \frac{1}{6} \cdot \frac{4}{5} = \frac{4}{30}$$

To make the numbers a little smaller before you go on, notice that you can reduce the fraction:

$$\text{Doreen} = \frac{2}{15}$$

Now you know how much Jerry and Doreen both ate, so you can add these amounts together:

$$\text{Jerry} + \text{Doreen} = \frac{1}{5} + \frac{2}{15}$$

Solve this problem as I show you in Chapter 10:

$$= \frac{3}{15} + \frac{2}{15} = \frac{5}{15}$$

This fraction reduces to ⅓. Now you know that Jerry and Doreen ate ⅓ of the pie, but the problem asks you how much is left. So finish up with some subtraction and write the answer:

$$1 - \frac{1}{3} = \frac{2}{3}$$

The amount of pie left over was ⅔.

Multiplying Decimals and Percents in Word Problems

In the preceding section, "Problems about Multiplying Fractions," I show you how the word *of* in a fraction word problem usually means *multiplication*. This idea is also true in word problems involving decimals and percents. The method for solving these two types of problems is similar, so I lump them together in this section.

You can easily solve word problems involving percents by changing the percents into decimals (see Chapter 12 for details). Here are a few common percents and their decimal equivalents:

$$25\% = 0.25 \qquad 50\% = 0.5 \qquad 75\% = 0.75 \qquad 99\% = 0.99$$

To the end: Figuring out how much money is left

One common type of problem gives you a starting amount — and a bunch of other information — and then asks you to figure out how much you end up with. Here's an example:

Maria's grandparents gave her $125 for her birthday. She put 40% of the money in the bank, spent 35% of what was left on a pair of shoes, and then spent the rest on a dress. How much did the dress cost?

Start at the beginning, forming a word equation to find out how much money Maria put in the bank:

Money in bank = 40% of $125

To solve this word equation, change the percent to a decimal and the word *of* to a multiplication sign; then multiply:

Money in bank = 0.4 · $125 = $50

Pay special attention to whether you're calculating how much of something was used up or how much of something is left over. If you need to work with the portion that remains, you may have to subtract the amount used from the amount you started with.

Because Maria started with $125, she had $75 left to spend:

Money left to spend = Money from grandparents   Money in bank

= $125 – $50 = $75

The problem then says that she spent 35% of *this* amount on a pair of shoes. Again, change the percent to a decimal and the word *of* to a multiplication sign:

Shoes = 35% of $75 = 0.35 · $75 = $26.25

She spent the rest of the money on a dress, so

Dress = $75 – $26.25 = $48.75

Therefore, Maria spent $48.75 on the dress.

Finding out how much you started with

Some problems give you the amount that you end up with and ask you to find out how much you started with. In general, these problems are harder because you're not used to thinking backwards. Here's an example, and it's kind of a tough one, so fasten your seat belt:

Maria received some birthday money from her aunt. She put her usual 40% in the bank, spent 75% of the rest on a purse, and when she was done, she had $12 left to spend on dinner. How much did her aunt give her?

This problem is similar to the one in the preceding section, but you need to start at the end and work backwards. Notice that the only dollar amount in the problem comes *after* the two percent amounts. The problem tells you that she *ends up with* $12 after two transactions — putting money in the bank and buying a purse — and asks you to find out how much you started with.

To solve this problem, set up two word equations to describe the two transactions:

> Money from aunt – money for bank = money after bank
>
> Money after bank – money for purse = $12

Notice what these two word equations are saying. The first tells you that Maria took the money from her aunt, subtracted some money to put in the bank, and left the bank with a new amount of money, which I'm calling *money after bank*. The second word equation starts where the first leaves off. It tells you that Maria took the money left over from the bank, subtracted some money for a purse, and ended up with $12.

This second equation already has an amount of money filled in, so start here. To solve this problem, realize that Maria spent 75% of her money *at that time* on the purse — that is, 75% of the money she still had after the bank:

> Money after bank – 75% of money after bank = $12

I'm going to make one small change to this equation so you can see what it's really saying:

> 100% of money after bank – 75% of money after bank = $12

Adding *100% of* doesn't change the equation because it really just means you're multiplying by 1. In fact, you can slip these two words in anywhere without changing what you mean, though you may sound ridiculous saying "Last night, I drove 100% of my car home from work, walked 100% of my dog, then took 100% of my wife to see 100% of a movie."

In this particular case, however, these words help you to make a connection because 100% – 75% = 25%; here's an even better way to write this equation:

> 25% · money after bank = $12

Before moving on, make sure you understand the steps that have brought you here. The percent circle (see Chapter 12) tells you that $12 ÷ 0.25 gives you the remaining amount of $48.

Okay, you know how much money Maria had after the bank, so you can plug this number into the first equation:

Money from aunt – money for bank = $48

Now you can use the same type of thinking to solve this equation (and it goes a lot more quickly this time!). First, Maria placed 40% of the money from her aunt in the bank:

Money from aunt – 40% of money from aunt = $48

Again, rewrite this equation to make what it's saying clearer:

100% of money from aunt – 40% of money from aunt = $48

Now, because 100% – 40% = 60%, rewrite it again.

60% · money from aunt = $48

At this point, you can use the percent circle, which I show you in Chapter 12, to solve the equation (see Figure 13-1). In this case, the percent circle tells you that $48 ÷ 0.6 gives you the remaining amount of $80. So Maria's aunt gave her $80 for her birthday.

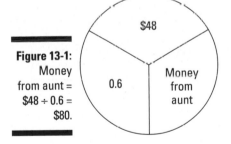

Figure 13-1:
Money
from aunt =
$48 ÷ 0.6 =
$80.

Handling Percent Increases and Decreases in Word Problems

Word problems that involve increasing or decreasing by a percentage add a final spin to percent problems. Typical percent-increase problems involve calculating the amount of a salary plus a raise, the cost of merchandise plus tax, or an amount of money plus interest or dividend. Typical percent decrease problems involve the amount of a salary minus taxes or the cost of merchandise minus a discount.

To tell you the truth, you may have already solved problems of this kind earlier in "Multiplying Decimals and Percents in Word Problems." But people often get thrown by the language of these problems — which, by the way, is the language of business — so I want to give you some practice in solving them.

Raking in the dough: Finding salary increases

A little street smarts should tell you that the words *salary increase* or *raise* means more money, so get ready to do some addition. Here's an example:

> Alison's salary was $40,000 last year, and at the end of the year she received a 5% raise. What will she earn this year?

To solve this problem, first realize that Alison got a raise. So whatever she makes this year, it will be *more* than she made last year. The key to setting up this type of problem is to think of percent increase as "100% of last year's salary plus 5% of last year's salary." Here's the word equation:

> This year's salary = 100% of last year's salary + 5% of last year's salary

---

A quick look at the distributive property

The *distributive property* says that multiplying a number by the sum of two other numbers in parentheses is the same thing as multiplying by the numbers in parentheses individually and then adding their products. In other words,

$$3(1 + 5) = 3(1) + 3(5)$$

If you evaluate both sides of the equation, here's what you get. You can see that they're equal:

$$3(6) = 3 + 15$$

$$18 = 18$$

The property also works for subtraction:

$$5(6 - 4) = 5(6) - 5(4)$$

$$5(2) = 30 - 20$$

$$10 = 10$$

In percent terms, you can say that the following statements are true:

$$\text{Salary } (100\% + 5\%) = \text{Salary } (100\%) + \text{Salary } (5\%)$$

$$\text{Salary } (100\% - 5\%) = \text{Salary } (100\%) - \text{Salary } (5\%)$$

What's the point? Well, you can solve percent problems either way, but adjusting the percents first is often easier. You may be able to do the addition or subtraction in your head, and as soon as you do the multiplication, you have your answer.

See Chapter 4 for more on the distributive property.

Now you can just add the percentages (see the nearby sidebar for why this works):

This year's salary = (100% + 5%) of last year's salary

= 105% of last year's salary

Change the percent to a decimal and the word *of* to a multiplication sign; then fill in the amount of last year's salary:

This year's salary = 1.05 · $40,000

Now you're ready to multiply:

This year's salary – $42,000

So Alison's new salary is $42,000.

Earning interest on top of interest

The word *interest* means more money. When you receive interest from the bank, you get more money. And when you pay interest on a loan, you pay more money. Sometimes people earn interest on the interest they earned earlier, which makes the dollar amounts grow even faster. Here's an example:

Bethany placed $9,500 in a one-year CD that paid 4% interest. The next year, she rolled this over into a bond that paid 6% per year. How much did Bethany earn on her investment in those two years?

This problem involves interest, so it's another problem in percent increase, only this time, you have to deal with two transactions. Take them one at a time.

The first transaction is a percent increase of 4% on $9,500. The following word equation makes sense:

Money after first year = 100% of initial deposit + 4% of initial deposit

Money after first year = (100% + 4%) of $9,500 = 104% of $9,500

Now change the percent to a decimal and the word *of* to a multiplication sign:

Money after first year = 1.04 · $9,500

Multiplication gives you this result:

Money after first year = $9,880

At this point, you're ready for the second transaction. This is a percent increase of 6% on $9,880:

Final amount = 106% of $9,880

Again, change the percent to a decimal and the word *of* to a multiplication sign:

Final amount = 1.06 · $9,880 = $10,472.80

Then subtract the initial deposit from the final amount:

Earnings = final amount − initial deposit = $10,472.80 − $9,500 = $972.80

So Bethany earned $972.80 on her investment.

Getting a deal: Calculating discounts

When you hear the words *discount* or *sale price,* think of subtraction. Here's an example:

> Greg has his eye on a television with a listed price of $2,100. The salesman offers him a 30% discount if he buys it today. What will the television cost with the discount?

In this problem, you need to realize that the discount lowers the price of the television, so you have to subtract:

Sale price = 100% of regular price − 30% of regular price

Now subtract percentages:

Sale price = (100% − 30%) of regular price = 70% of regular price

At this point, you can fill in the details, as I show you throughout this chapter:

Sale price = 0.7 · $2,100 = $1,470

Thus, the television will cost $1,470 with the discount.

Part IV
Picturing and Measuring — Graphs, Measures, Stats, and Sets

The 5th Wave   By Rich Tennant

"He ran some linear equations, threw in a few theorems, and before I knew it I was buying rustproofing."

In this part . . .

I introduce you to scientific notation as a convenient way to represent very large and very small numbers. I discuss two important systems of measurement — the English system and the metric system — and show you how to convert from one to the other. You do some geometry and discover how to find the perimeter and area of important shapes such as squares, rectangles, and triangles. I provide some quick information on how to read a variety of graphs and present the Cartesian graph. You get started with some basic set theory, statistics, and probability. Finally, I show you how to apply these skills to solve word problems involving measurement and geometry.

Chapter 14

A Perfect Ten: Condensing Numbers with Scientific Notation

In This Chapter

▶ Knowing how to express powers of ten in exponential form

▶ Appreciating how and why scientific notation works

▶ Understanding order of magnitude

▶ Multiplying numbers in scientific notation

Scientists often work with very small or very large measurements — the distance to the next galaxy, the size of an atom, the mass of the Earth, or the number of bacteria cells growing in last week's leftover Chinese takeout. To save on time and space — and to make calculations easier — people developed a sort of shorthand called scientific notation.

Scientific notation uses a sequence of numbers known as the *powers of ten,* which I introduce in Chapter 2:

1   10   100   1,000   10,000   100,000   1,000,000   10,000,000 ...

Each number in the sequence is 10 times more than the preceding number.

Powers of ten are easy to work with, especially when you're multiplying and dividing, because you can just add or drop zeros or move the decimal point. They're also easy to represent in exponential form (as I show you in Chapter 4):

10^0 10^1 10^2 10^3 10^4 10^5 10^6 10^7 ...

Scientific notation is a handy system for writing very large and very small numbers without writing a bunch of 0s. It uses both decimals and exponents (so if you need a little brushing up on decimals, flip to Chapter 11). In this chapter, I introduce you to this powerful method of writing numbers. I also explain the order of magnitude of a number. Finally, I show you how to multiply numbers written in scientific notation.

First Things First: Powers of Ten as Exponents

Scientific notation uses powers of ten expressed as exponents, so you need a little background before you can jump in. In this section, I round out your knowledge of exponents, which I first introduce in Chapter 4.

Counting up zeros and writing exponents

Numbers starting with a 1 and followed by only 0s (such 10, 100, 1,000, 10,000, and so forth) are called *powers of ten,* and they're easy to represent as exponents. Powers of ten are the result of multiplying 10 times itself any number of times.

To represent a number that's a power of 10 as an exponential number, count the zeros and raise 10 to that exponent. For example, 1,000 has three zeros, so $1,000 = 10^3$ (10^3 means to take 10 times itself three times, so it equals $10 \cdot 10 \cdot 10$). Table 14-1 shows a list of some powers of ten.

Table 14-1	Powers of Ten Expressed as Exponents
Number	*Exponent*
1	10^0
10	10^1
100	10^2
1,000	10^3
10,000	10^4
100,000	10^5
1,000,000	10^6

After you know this trick, representing a lot of large numbers as powers of ten is easy — just count the 0s! For example, the number 1 trillion — 1,000,000,000,000 — is a 1 with twelve 0s after it, so

$$1,000,000,000,000 = 10^{12}$$

This trick may not seem like a big deal, but the higher the numbers get, the more space you save by using exponents. For example, a really big number is a *googol*, which is 1 followed by a hundred 0s. You can write this:

10,000,000,000,000,000,000,000,000,000,000,000,000,000,000,000,000,000,000,
000,000,000,000,000,000,000,000,000,000,000,000,000,000,000,000

As you can see, a number of this size is practically unmanageable. You can save yourself some trouble and write 10^{100}.

A 10 raised to a negative number is also a power of ten.

You can also represent decimals using *negative exponents*. For example,

$$10^{-1} = 0.1 \qquad 10^{-2} = 0.01 \qquad 10^{-3} = 0.001 \qquad 10^{-4} = 0.0001$$

Although the idea of negative exponents may seem strange, it makes sense when you think about it alongside what you know about positive exponents. For example, to find the value of 10^7, start with 1 and make it *larger* by moving the decimal point 7 spaces to the *right*:

$$10^7 = 10,000,000$$

Similarly, to find the value of 10^{-7}, start with 1 and make it *smaller* by moving the decimal point 7 spaces to the *left*:

$$10^{-7} = 0.0000001$$

Negative powers of 10 always have *one fewer 0* between the 1 and the decimal point than the power indicates. In this example, notice that 10^{-7} has six 0s between them.

As with very large numbers, using exponents to represent very small decimals makes practical sense. For example,

$$10^{-23} = 0.00000000000000000000001$$

As you can see, this decimal is easy to work with in its exponential form but almost impossible to read otherwise.

Adding exponents to multiply

An advantage of using the exponential form to represent powers of ten is that this form is a cinch to multiply. To multiply two powers of ten in exponential form, add their exponents. Here are a few examples:

✓ $10^1 \cdot 10^2 = 10^{1+2} = 10^3$

Here, I simply multiply these numbers:

$10 \cdot 100 = 1,000$

✓ $10^{14} \cdot 10^{15} = 10^{14+15} = 10^{29}$

Here's what I'm multiplying:

$100,000,000,000,000 \cdot 1,000,000,000,000,000$

$= 100,000,000,000,000,000,000,000,000,000$

You can verify that this multiplication is correct by counting the 0s.

✓ $10^{100} \cdot 10^0 = 10^{100+0} = 10^{100}$

Here I'm multiplying a googol by 1 (any number raised to an exponent of 0 equals 1), so the result is a googol.

In each of these cases, you can think of multiplying powers of ten as adding extra 0s to the number.

The rules for multiplying powers of ten by adding exponents also applies to negative exponents. For example,

$10^3 \cdot 10^{-5} = 10^{(3-5)} = 10^{-2} = 0.01$

Working with Scientific Notation

Scientific notation is a system for writing very large and very small numbers that makes them easier to work with. Every number can be written in scientific notation as the product of two numbers (two numbers multiplied together):

✓ A decimal greater than or equal to 1 and less than 10 (see Chapter 11 for more on decimals)

✓ A power of ten written as an exponent (see the preceding section)

Writing in scientific notation

Here's how to write any number in scientific notation:

1. **Write the number as a decimal (if it isn't one already).**

   Suppose you want to change the number 360,000,000 to scientific notation. First, write it as a decimal:

   360,000,000.0

2. **Move the decimal point just enough places to change this number to a new number that's between 1 and 10.**

   Move the decimal point to the right or left so that only one nonzero digit comes before the decimal point. Drop any leading or trailing zeros as necessary.

   Using 360,000,000.0, only the 3 should come before the decimal point. So move the decimal point eight places to the left, drop the trailing zeros, and get 3.6:

   360,000,000.0 becomes 3.6

3. **Multiply the new number by 10 raised to the number of places you moved the decimal point in Step 2.**

   You moved the decimal point eight places, so multiply the new number by 10^8:

   $3.6 \cdot 10^8$

4. **If you moved the decimal point to the right in Step 2, put a minus sign on the exponent.**

   You moved the decimal point to the left, so you don't have to take any action here. Thus, 360,000,000 in scientific notation is $3.6 \cdot 10^8$.

Changing a decimal to scientific notation basically follows the same process. For example, suppose you want to change the number 0.00006113 to scientific notation:

1. **Write 0.00006113 as a decimal (this step's easy, because it's already a decimal):**

   0.00006113

2. **To change 0.00006113 to a new number between 1 and 10, move the decimal point five places to the right and drop the leading zeros:**

   6.113

3. **Because you moved the decimal point five places, multiply the new number by 10^5:**

   $6.113 \cdot 10^5$

4. **Because you moved the decimal point to the right, put a minus sign on the exponent:**

   $6.113 \cdot 10^{-5}$

   So 0.00006113 in scientific notation is $6.113 \cdot 10^{-5}$.

After you get used to writing numbers in scientific notation, you can do it all in one step. Here are a few examples:

$$17,400 = 1.7400 \cdot 10^4$$
$$212.04 = 2.1204 \cdot 10^2$$
$$0.003002 = 3.002 \cdot 10^{-3}$$

Why scientific notation works

After you understand *how* scientific notation works, you're in a better position to understand *why* it works. Suppose you're working with the number 4,500. First of all, you can multiply any number by 1 without changing it, so here's a valid equation:

$$4,500 = 4,500 \cdot 1$$

Because 4,500 ends in a 0, it's divisible by 10 (see Chapter 7 for info on divisibility). So you can factor out a 10 as follows:

$$4,500 = 450 \cdot 10$$

Also, because 4,500 ends in two 0s, it's divisible by 100, so you can factor out 100:

$$4,500 = 45 \cdot 100$$

In each case, you drop another 0 after the 45 and place it after the 1. At this point, you have no more 0s to drop, but you can continue the pattern by moving the decimal point one place to the left:

$$4,500 = 4.5 \cdot 1,000$$
$$4,500 = 0.45 \cdot 10,000$$
$$4,500 = 0.045 \cdot 100,000$$

What you've been doing from the beginning is moving the decimal point one place to the left and multiplying by 10. But you can just as easily move the decimal point one place to the right and multiply by 0.1, two places right by multiplying by 0.01, and three places right by multiplying by 0.001:

$$4,500 = 45,000 \cdot 0.1$$

$$4,500 = 450,000 \cdot 0.01$$

$$4,500 = 4,500,000 \cdot 0.001$$

As you can see, you have total flexibility to express 4,500 as a decimal multiplied by a power of ten. As it happens, in scientific notation, the decimal must be between 1 and 10, so the following form is the equation of choice:

$$4,500 = 4.5 \cdot 1,000$$

The final step is to change 1,000 to exponential form. Just count the 0s in 1,000 and write that number as the exponent on the 10:

$$4,500 = 4.5 \cdot 10^3$$

The net effect is that you moved the decimal point three places to the left and raised 10 to an exponent of 3. You can see how this idea can work for any number, no matter how large or small.

Understanding order of magnitude

A good question to ask is why scientific notation always uses a decimal between 1 and 10. The answer has to do with order of magnitude. Order of magnitude is a simple way to keep track of roughly how large a number is so you can compare numbers more easily. The *order of magnitude* of a number is its exponent in scientific notation. For example,

$703 = 7.03 \cdot 10^2$ — order of magnitude is 2

$600,000 = 6 \cdot 10^5$ — order of magnitude is 5

$0.00095 = 9.5 \cdot 10^{-4}$ — order of magnitude is −4

Every number starting with 10 but less than 100 has an order of magnitude of 1. Every number starting with 100 but less than 1,000 has an order of magnitude of 2.

Multiplying with scientific notation

Multiplying numbers that are in scientific notation is fairly simple because multiplying powers of ten is easy, as you see earlier in this chapter in "Adding exponents to multiply." Here's how to multiply two numbers that are in scientific notation:

1. **Multiply the two decimal parts of the numbers.**

   Suppose you want to multiply the following:

   $$(4.3 \cdot 10^5)(2 \cdot 10^7)$$

   Multiplication is commutative (see Chapter 4), so you can change the order of the numbers without changing the result. And because of the associative property, you can also change how you group the numbers. Therefore, you can rewrite this problem as

   $$(4.3 \cdot 2)(10^5 \cdot 10^7)$$

   Multiply what's in the first set of parentheses — $4.3 \cdot 2$ — to find the decimal part of the solution:

   $$4.3 \cdot 2 = 8.6$$

2. **Multiply the two exponential parts by adding their exponents.**

   Now multiply 10^5 by 10^7:

   $$10^5 \cdot 10^7 = 10^{5+7} = 10^{12}$$

3. **Write the answer as the product of the numbers you found in Steps 1 and 2.**

   $$8.6 \cdot 10^{12}$$

4. **If the decimal part of the solution is 10 or greater, move the decimal point one place to the left and add 1 to the exponent.**

   Because 8.6 is less than 10, you don't have to move the decimal point again, so the answer is $8.6 \cdot 10^{12}$.

   Note: This number equals 8,600,000,000,000.

Because scientific notation uses positive decimals less than 10, when you multiply two of these decimals, the result is always a positive number less than 100. So in Step 4, you never have to move the decimal point more than one place to the left.

This method even works when one or both of the exponents are negative numbers. For example, suppose you want to multiply the following:

$$(6.02 \cdot 10^{23})(9 \cdot 10^{-28})$$

1. **Multiply 6.02 by 9 to find the decimal part of the answer:**

   $6.02 \cdot 9 = 54.18$

2. **Multiply 10^{23} by 10^{-28} by adding the exponents (check Chapter 4 if you need info on adding negative numbers):**

   $10^{23} \cdot 10^{-28} = 10^{23 + -28} = 10^{-5}$

3. **Write the answer as the product of the two numbers:**

   $54.18 \cdot 10^{-5}$

4. **Because 54.18 is greater than 10, move the decimal point one place to the left and add 1 to the exponent:**

   $5.418 \cdot 10^{-4}$

Note: In decimal form, this number equals 0.0005418.

Scientific notation really pays off when you're multiplying very large and very small numbers. If you'd tried to multiply the numbers in the preceding example the usual way, here's what you would've been up against:

$602,000,000,000,000,000,000,000 \cdot 0.00000000000000000000000000009$

As you can see, scientific notation makes the job a lot easier.

Chapter 15

How Much Have You Got? Weights and Measures

*I*n Chapter 4, I introduce you to *units,* which are things that can be counted, such as apples, coins, or hats. Apples, coins, and hats are easy to count because they're *discrete* — that is, you can easily see where one ends and the next one begins. But not everything is so easy. For example, how do you count water — by the drop? Even if you tried, exactly how big is a drop?

That's where units of measurement come in. A *unit of measurement* allows you to count something that isn't discrete: an amount of a liquid or solid, the distance from one place to another, a length of time, the speed at which you're traveling, or the temperature of the air.

In this chapter, I discuss two important systems of measurement: English and metric. You're probably familiar with the English system already, and you may know more than you think about the metric system. Each of these measurement systems provides a different way to measure distance, volume, weight (or mass), time, and speed. Next, I show you how to estimate metric amounts in English units. Finally, I show how to convert from English units to metric and vice versa.

Examining Differences between the English and Metric Systems

The two most common measurement systems today are the *English system* and the *metric system.*

Most Americans learn the units of the English system — for example, pounds and ounces, feet and inches, and so forth — and use them every day. Unfortunately, the English system is awkward for use with math. English units such as inches and fluid ounces are often measured in fractions, which (as you know from Chapters 9 and 10) can be difficult to work with.

The *metric system* was invented to simplify the application of math to measurement. Metric units are based on the number 10, which makes them much easier to work with. Parts of units are expressed as decimals, which (as Chapter 11 shows you) are much friendlier than fractions.

Yet despite these advantages, the metric system has been slow to catch on in the U.S. Many Americans feel comfortable with English units and reluctant to part with them. This is understandable. For example, if I ask, you to carry a 20 lb. bag for 1/4 of a mile, you know what to expect. However, if I ask you to carry a bag weighing 10 kilograms half a kilometer, you may not be sure.

In this section, I show you the basic units of measurement for both the English and metric systems.

If you want an example of the importance of converting carefully, you may want to look to NASA — they kind of lost a Mars orbiter in the late 1990s because an engineering team used English units and NASA used metric to navigate!

Looking at the English system

The *English system of measurement* is most commonly used in the United States (but, ironically, not in England). Although you're probably familiar with most of the English units of measurement, in the following list, I make sure you know the most important ones. I also show you some equivalent values that can help you do conversions from one type of unit to another.

- ✔ **Units of distance:** Distance — also called length — is measured in inches (in.), feet (ft.), yards (yd.), and miles (mi.):

    12 inches = 1 foot

    3 feet = 1 yard

    5,280 feet = 1 mile

✔ **Units of fluid volume:** Fluid volume (also called *capacity*) is the amount of space occupied by a liquid such as water, milk, or wine. I discuss volume when I talk about geometry in Chapter 16. Volume is measured in fluid ounces (fl. oz.), cups (c.), pints (pt.), quarts (qt.), and gallons (gal.):

8 fluid ounces = 1 cup

2 cups = 1 pint

2 pints = 1 quart

4 quarts = 1 gallon

Units of fluid volume are typically used for measuring the volume of things that can be poured. The volume of solid objects is more commonly measured in cubic units of distance such as cubic inches, cubic feet, and so forth.

✔ **Units of weight:** Weight is the measurement of how strongly gravity pulls an object toward Earth. Weight is measured in ounces (oz.), pounds (lb.), and tons.

16 ounces = 1 pound

2,000 pounds = 1 ton

Don't confuse *fluid ounces,* which measure volume, with *ounces,* which measure weight. These units are two completely different types of measurements!

✔ **Units of time:** Time is hard to define, but everybody knows what it is. Time is measured in seconds, minutes, hours, days, weeks, and years:

60 seconds = 1 minute

60 minutes = 1 hour

24 hours = 1 day

7 days = 1 week

365 days ≈ 1 year

The conversion from days to years is approximate, because earth's daily rotation on its axis and its yearly revolution around the sun aren't exactly synchronized. A year is closer to 365.25 days, which is why leap years exist.

I left months out of the picture because the definition of a month is imprecise — it can vary from 28 to 31 days.

✔ **Unit of speed:** Speed is the measurement of how much time an object takes to move a given distance. The most common unit of speed is miles per hour (mph).

✔ **Unit of temperature:** Temperature measures how much heat an object contains. This object can be a glass of water, a turkey in the oven, or the air surrounding your house. Temperature is measured in degrees Fahrenheit (°F).

Looking at the metric system

Like the English system, the metric system provides units of measurement for distance, volume, and so on. Unlike the English system, however, the metric system builds these units using a *basic unit* and a set of *prefixes*.

Table 15-1 shows five important basic units in the metric system.

Table 15-1	Five Basic Metric Units
Measure Of	*Basic Metric Unit*
Distance	Meter
Volume (capacity)	Liter
Mass (weight)	Gram
Time	Second
Temperature	Degree (Celsius)

For scientific purposes, the metric system has been updated to more rigorously defined *System of International Units* (SI). Each basic SI unit correlates directly to a measurable scientific process that defines it. In SI, the kilogram (not the gram) is the basic unit of mass, the kelvin is the basic unit of temperature, and the liter is not considered a basic unit. For technical reasons, scientists tend to use the more rigidly defined SI, but most other people use the looser metric system. In everyday practice, you can think of the units listed in Table 15-1 as basic units.

Table 15-2 shows ten metric prefixes, with the three most commonly used in bold (see Chapter 14 for more information on powers of ten).

Table 15-2		Ten Metric Prefixes	
Prefix	*Meaning*	*Number*	*Power of Ten*
Giga-	One billion	1,000,000,000	10^9
Mega-	One million	1,000,000	10^6
Kilo-	*One thousand*	*1,000*	*10^3*
Hecta-	One hundred	100	10^2
Deca-	Ten	10	10^1
(none)	One	1	10^0
Deci-	One tenth	0.1	10^{-1}
Centi-	*One hundredth*	*0.01*	*10^{-2}*
Milli-	*One thousandth*	*0.001*	*10^{-3}*
Micro-	One millionth	0.000001	10^{-6}
Nano-	One billionth	0.000000001	10^{-9}

Large and small metric units are formed by linking a basic unit with a prefix. For example, linking the prefix *kilo-* to the basic unit *meter* gives you the *kilometer,* which means 1,000 meters. Similarly, linking the prefix *milli-* to the basic unit *liter* gives you the *milliliter,* which means 0.001 (one thousandth) of a meter.

Here's a list giving you the basics:

✔ **Units of distance:** The basic metric unit of distance is the *meter* (m). Other common units are millimeters (mm), centimeters (cm), and kilometers (km):

> 1 kilometer = 1,000 meters
>
> 1 meter = 100 centimeters
>
> 1 meter = 1,000 millimeters

✔ **Units of fluid volume:** The basic metric unit of fluid volume (also called capacity) is the *liter* (L). Another common unit is the milliliter (mL):

> 1 liter = 1,000 milliliters

Note: One milliliter is equal to 1 cubic centimeter (cc).

✔ **Units of mass:** Technically speaking, the metric system measures not weight but mass. *Weight* is the measurement of how strongly gravity pulls an object toward Earth. *Mass,* however, is the measurement of the amount of matter an object has. If you traveled to the moon, your weight would change, so you would feel lighter. But your mass would remain the same, so all of you would still be there. Unless you're planning a trip into outer space or performing a scientific experiment, you probably don't need to know the difference between weight and mass. In this chapter, you can think of them as equivalent, and I use the word *weight* when referring to metric mass.

The basic unit of weight in the metric system is the *gram* (g). Even more commonly used, however, is the kilogram (kg):

> 1 kilogram = 1,000 grams

Note: 1 kilogram of water has a volume of 1 liter.

✔ **Units of time:** As in the English system, the basic metric unit of time is a *second* (s). For most purposes, people also use other English units such as minutes, hours, and so forth.

For many scientific purposes, the second is the only unit used to measure time. Large numbers of seconds and small fractions of sections are represented with *scientific notation,* which I cover in Chapter 14.

✔ **Units of speed:** For most purposes, the most common metric unit of speed (also called velocity) is *kilometers per hour* (km/hr). Another common unit is *meters per second* (m/s).

✔ **Units of temperature (degrees Celsius or Centigrade):** The basic metric unit of temperature is the *Celsius degree* (°C), also called the *Centigrade degree*. The Celsius scale is set up so that at sea level, water freezes at 0°C and boils at 100°C.

Scientists often use another unit — the Kelvin (K) — to talk about temperature. The degrees are the same size as in Celsius, but 0 K is set at *absolute zero,* the temperature at which atoms don't move at all. Absolute zero is approximately equal to –273.15°C.

Estimating and Converting between the English and Metric Systems

Most Americans use the English system of measurement all the time but have only a passing acquaintance with the metric system. But metric units are being used more commonly as the units for tools, footraces, soft drinks, and many other things. Also, if you travel abroad, you need to know how far 100 kilometers is or how long you can drive on 10 liters of gasoline.

In this section, I show you how to make ballpark estimates of metric units in terms of English units, which can help you feel more comfortable with metric units. I also show you how to convert between English and metric units, which is a common type of math problem.

When I talk about *estimating*, I mean very loose ways of measuring metric amounts using the English units you are familiar with. In contrast, when I talk about *converting*, I mean using an equation to change from one system of units to the other. Neither method is exact, but converting provides a much closer approximation (and takes longer) than estimating.

Estimating in the metric system

One reason people sometimes feel uncomfortable using the metric system is that when you're not familiar with it, estimating amounts in practical terms is hard. For example, if I tell you that we're going out to a beach that's ¼ mile away, you prepare yourself for a short walk. And if I tell you that it's 10 miles away, you head for the car. But what do you do with the information that the beach is 3 kilometers away?

Similarly, if I tell you that the temperature is 85 degrees Fahrenheit, you'll probably wear a bathing suit or shorts. And if I tell you it's 40 degrees Fahrenheit, you'll probably wear a coat. But what do you wear if I tell you that the temperature is 25 degrees Celsius?

In this section, I give you a few rules of thumb to estimate metric amounts. In each case, I show you how a common metric unit compares with an English unit that you already feel comfortable with.

Approximating short distances: 1 meter is about 1 yard (3 feet)

Here's how to convert meters to feet: 1 meter ≈ 3.26 feet. But for estimating, use the simple rule that 1 meter is about 1 yard (that is, about 3 feet).

By this estimate, a 6-foot man stands about 2 meters tall. A 15-foot room is 5 meters wide. And a football field that's 100 yards long is about 100 meters long. Similarly, a river with a depth of 4 meters is about 12 feet deep. A mountain that's 3,000 meters tall is about 9,000 feet. And a child who is only half a meter tall is about a foot and a half.

Estimating longer distances and speed

Here's how to convert kilometers to miles: 1 kilometer ≈ 0.62 miles. For a ball-park estimate, you can remember that 1 kilometer is about ½ a mile. By the same token, 1 kilometer per hour is about ½ mile per hour.

This guideline tells you that if you live 2 miles from the nearest supermarket, then you live about 4 kilometers from there. A marathon of 26 miles is about 52 kilometers. And if you run on a treadmill at 6 miles per hour, then you can run at about 12 kilometers per hour. By the same token, a 10 kilometer race is about 5 miles. If the Tour de France is about 4,000 kilometers, then it's about 2,000 miles. And if light travels about 300,000 kilometers per second, then it travels about 150,000 miles per second.

Approximating volume: 1 liter is about 1 quart (¼ gallon)

Here's how to convert liters to gallons: 1 liter ≈ 0.26 gallons. A good estimate here is that 1 liter is about 1 quart (that is, there are about 4 liters to the gallon).

Using this estimate, a gallon of milk is 4 quarts, so it's about 4 liters. If you put 10 gallons of gasoline in your tank, this is about 40 liters. In the other direction, if you buy a 2-liter bottle of cola, you have about 2 quarts. If you buy an aquarium with a 100-liter capacity, it holds about 25 gallons of water. And if a pool holds 8,000 liters of water, it holds about 2,000 gallons.

Estimating weight: 1 kilogram is about 2 pounds

Here's how to convert kilograms to pounds: 1 kilogram ≈ 2.20 pounds. For estimating, figure that 1 kilogram is equal to about 2 pounds.

By this estimate, a 5-kilogram bag of potatoes weighs about 10 pounds. If you can bench press 70 kilograms, then you can bench press about 140 pounds. And because a liter of water weighs exactly 1 kilogram, you know that a quart of water weighs about 2 pounds. Similarly, if a baby weighs 8 pounds at birth, he or she weighs about 4 kilograms. If you weigh 150 pounds, then you weigh about 75 kilograms. And if your New Year's resolution is to lose 20 pounds, then you want to lose about 10 kilograms.

Estimating temperature

The most common reason for estimating temperature in Celsius is in connection with the weather. The formula for converting from Celsius to Fahrenheit is kind of messy:

Fahrenheit = Celsius · ⅘ + 32

Instead, use the handy chart in Table 15-3.

Table 15-3	Comparing Celsius and Fahrenheit Temperatures	
Celsius (Centigrade)	*Description*	*Fahrenheit*
0 degrees	Cold	32 degrees
10 degrees	Cool	50 degrees
20 degrees	Warm	68 degrees
30 degrees	Hot	86 degrees

Any temperature below 0°C is cold, and any temperature over 30°C is hot. Most of the time, the temperature falls in this middling range. So now you know that when the temperature is 6°C, you want to wear a coat. When it's 14°C, you may want a sweater or at least long sleeves. And when it's 25°C, head for the beach!

Converting units of measurement

Many books give you one formula for converting from English to metric and another for converting from metric to English. People often find this conversion method confusing because they have trouble remembering which formula to use in which direction.

In this section, I show you a simple way to convert between English and metric units that uses only one formula for each type of conversion.

Understanding conversion factors

When you multiply any number by 1, that number stays the same. For example, $36 \cdot 1 = 36$. And when a fraction has the same *numerator* (top number) and *denominator* (bottom number), that fraction equals 1 (see Chapter 10 for details). So when you multiply a number by a fraction that equals 1, the number stays the same. For example:

$$36 \cdot \frac{5}{5} = 36$$

If you multiply a measurement by a special fraction that equals 1, you can switch from one unit of measurement to another without changing the value. People call such fractions *conversion factors*.

Take a look at some equations that show how metric and English units are related (all conversions between English and metric units are approximate):

- 1 meter ≈ 3.26 feet
- 1 kilometer ≈ 0.62 miles
- 1 liter ≈ 0.26 gallons
- 1 kilogram ≈ 2.20 pounds

Because the values on each side of the equations are equal, you can create fractions that are equal to 1, such as

- $\dfrac{1 \text{ meter}}{3.26 \text{ feet}}$ or $\dfrac{3.26 \text{ feet}}{1 \text{ meter}}$

- $\dfrac{1 \text{ kilometer}}{0.62 \text{ miles}}$ or $\dfrac{0.62 \text{ miles}}{1 \text{ kilometer}}$

- $\dfrac{1 \text{ liter}}{0.26 \text{ gallons}}$ or $\dfrac{0.26 \text{ gallons}}{1 \text{ liter}}$

- $\dfrac{1 \text{ kilogram}}{2.2 \text{ pounds}}$ or $\dfrac{2.2 \text{ pounds}}{1 \text{ kilogram}}$

After you understand how units of measurement cancel (which I discuss in the next section), you can easily choose which fractions to use to switch between units of measurement.

Canceling units of measurement

When you're multiplying fractions, you can cancel any factor that appears in both the numerator and denominator (see Chapter 9 for details). Just as with numbers, you can also cancel out units of measurement in fractions. For example, suppose you want to evaluate this fraction:

$$\frac{6 \text{ gallons}}{2 \text{ gallons}}$$

You already know that you can cancel out a factor of 2 in both the numerator and denominator. But you can also cancel out the unit *gallons* in both the numerator and denominator:

$$= \frac{\overset{3}{\cancel{6 \text{ gallons}}}}{\cancel{2 \text{ gallons}}}$$

So this fraction simplifies as follows:

$$= 3$$

Converting units

After you understand how to cancel out units in fractions and how to set up fractions equal to 1 (see the preceding sections), you have a foolproof system for converting units of measurement.

Suppose you want to convert 7 meters into feet. Using the equation 1 meter = 3.26 feet, you can make a fraction out of the two values as follows:

$$\frac{1 \text{ meter}}{3.26 \text{ feet}} = 1 \quad \text{or} \quad \frac{3.26 \text{ feet}}{1 \text{ meter}} = 1$$

Both fractions equal 1 because the numerator and the denominator are equal. So you can multiply the quantity you're trying to convert (7 meters) by one of these fractions without changing it. Remember that you want the *meters* unit to cancel out. You already have the word *meters* in the numerator (to make this clear, place 1 in the denominator), so use the fraction that puts *1 meter* in the denominator:

$$\frac{7 \text{ meters}}{1} \cdot \frac{3.26 \text{ feet}}{1 \text{ meter}}$$

Now cancel out the unit that appears in both the numerator and denominator:

$$= \frac{7 \ \cancel{\text{meters}}}{1} \cdot \frac{3.26 \text{ feet}}{1 \ \cancel{\text{meter}}}$$

At this point, the only value in the denominator is 1, so you can ignore it. And the only unit left is *feet*, so place it at the end of the expression:

$$= 7 \cdot 3.26 \text{ feet}$$

Now do the multiplication (Chapter 11 shows how to multiply decimals):

$$= 22.82 \text{ feet}$$

It may seem strange that the answer appears with the units already attached, but that's the beauty of this method: When you set up the right expression, the answer just appears.

You can get more practice converting units of measurement in Chapter 18, where I show you how to set up conversion chains and tackle word problems involving measurement.

Chapter 16

Picture This: Basic Geometry

In This Chapter

▶ Knowing the basic components of geometry: Points, lines, angles, and shapes

▶ Examining two-dimensional shapes

▶ Looking at solid geometry

▶ Finding out how to measure a variety of shapes

Geometry is the mathematics of figures such as squares, circles, triangles, lines, and so forth. Because geometry is the math of physical space, it's one of the most useful areas of math. Geometry comes into play when measuring rooms or walls in your house, the area of a circular garden, the volume of water in a pool, or the shortest distance across a rectangular field.

Although geometry is usually a year-long course in high school, you may be surprised at how quickly you can pick up what you need to know about basic geometry. Much of what you discover in a geometry course is how to write geometric proofs, which you don't need for algebra — or trigonometry, or even calculus.

In this chapter, I give you a quick and practical overview of geometry. First, I show you four important concepts in plane geometry: points, lines, angles, and shapes. Then I give you the basics on geometric shapes, from flat circles to solid cubes. Finally, I discuss how to measure geometric shapes by finding the area and perimeter of two-dimensional forms and the volume and surface area of some geometric solids.

Of course, if you want to know more about geometry, the ideal place to look beyond this chapter is *Geometry For Dummies* (published by Wiley)!

Getting on the Plane: Points, Lines, Angles, and Shapes

Plane geometry is the study of figures on a two-dimensional surface — that is, on a plane. You can think of the plane as a piece of paper with no thickness at all. Technically, a plane doesn't end at the edge of the paper — it continues forever.

In this section, I introduce you to four important concepts in plane geometry: points, lines, angles, and shapes (such as squares, circles, triangles, and so forth).

Making some points

A *point* is a location on a plane. It has no size or shape. Although in reality a point is too small to be seen, you can represent it visually in a drawing by using a dot.

When two lines intersect, as shown above, they share a single point. Additionally, each corner of a polygon is a point. (See below for more on lines and polygons.)

Knowing your lines

A *line* — also called a *straight line* — is pretty much what it sounds like; it marks the shortest distance between two points, but it extends infinitely in both directions. It has length but no width, making it a one-dimensional (1-D) figure.

Given any two points, you can draw exactly one line that passes through both of them. That is, two points *determine* a line.

When two lines intersect, they share a single point. When two lines don't intersect, they are *parallel,* which means that they remain the same distance from each other everywhere. A good visual aid for parallel lines is a set of railroad tracks. In geometry, you draw a line with arrows at both ends. Arrows on either end of a line mean that the line goes on forever (as you can see in Chapter 1, where I discuss the number line).

A *line segment* is a piece of a line that has endpoints, as shown here.

A *ray* is a piece of a line that starts at a point and extends infinitely in one direction, kind of like a laser. It has one endpoint and one arrow.

Figuring the angles

An *angle* is formed when two rays extend from the same point.

Angles are typically used in carpentry to measure the corners of objects. They're also used in navigation to indicate a sudden change in direction. For example, when you are driving, it's common to distinguish when the angle of a turn is "sharp" or "not so sharp."

The sharpness of an angle is usually measured in *degrees*. The most common angle is the *right angle* — the angle at the corner of a square — which is a 90° (90 degree) angle:

Angles that have fewer than 90° — that is, angles that are sharper than a right angle — are called *acute angles*, like this one:

Angles that measure greater than 90° — that is, angles that aren't as sharp as a right angle — are called *obtuse angles*, as seen here:

When an angle is exactly 180°, it forms a straight line and is called a *straight angle*.

Shaping things up

A shape is any closed geometrical figure that has an *inside* and an *outside*. Circles, squares, triangles, and larger polygons are all examples of shapes.

Much of plane geometry focuses on different types of shapes. In the next section, I show you how to identify a variety of shapes. Later in this chapter, I show you how to measure these shapes.

Closed Encounters: Shaping Up Your Understanding of 2-D Shapes

A *shape* is any closed two-dimensional (2-D) geometrical figure that has an *inside* and an *outside,* separated by the *perimeter* (boundary) of the shape. The *area* of a shape is the measurement of the size inside that shape.

A few shapes that you're probably familiar with include the square, rectangle, and triangle. However, many shapes don't have names, as you can see in Figure 16-1.

Figure 16-1:
Unnamed
shapes.

Measuring the perimeter and area of shapes is useful for a variety of applications, from land surveying (to get information about a parcel of land that you're measuring) to sewing (to figure out how much material you need for a project). In this section, I introduce you to a variety of geometric shapes. Later in the chapter, I show you how to find the perimeter and area of each, but for now I just acquaint you with them.

Circles

A circle is the set of all points that are a constant distance from the circle's center. The distance from any point on the circle to its center is called the *radius* of the circle. The distance from any point on the circle straight through the center to the other side of the circle is called the *diameter* of the circle.

Unlike the polygons I discuss later, a circle has no straight edges. The ancient Greeks — who invented much of geometry as we know it today — thought that the circle was the most perfect geometric shape.

Polygons

A *polygon* is any shape whose sides are all straight. Every polygon has three or more sides (if it had fewer than three, it wouldn't really be a shape at all). Following are a few of the most common polygons.

Triangles

The most basic shape with straight sides is the *triangle*, a three-sided polygon. You find out all about triangles when you study trigonometry (and what better place to begin than *Trigonometry For Dummies?*). Triangles are classified on the basis of their sides and angles. Take a look at the differences (and see Figure 16-2):

- ✔ **Equilateral:** An *equilateral triangle* has three sides that are all the same length and three angles that all measure 60 degrees.

- ✔ **Isosceles:** An *isosceles triangle* has two sides that are the same length and two equal angles.

- ✔ **Scalene:** *Scalene triangles* have three sides that are all different lengths and three angles that are all unequal.

- ✔ **Right:** A *right triangle* has one right angle. It may be isosceles or scalene.

Figure 16-2:
Types of
triangles.

Equilateral   Isosceles   Scalene   Right

Quadrilaterals

A *quadrilateral* is any shape that has four straight sides. Quadrilaterals are one of the most common shapes you see in daily life. If you doubt this, look around and notice that most rooms, doors, windows, and tabletops are quadrilaterals. Here, I introduce you to a few common quadrilaterals (Figure 16-3 shows you what they look like):

- ✔ **Square:** A *square* has four right angles and four sides of equal length; also, both pairs of opposite sides (sides directly across from each other) are parallel.

- ✔ **Rectangle:** Like a square, a *rectangle* has four right angles and two pairs of opposite sides that are parallel. Unlike the square, however, although opposite sides are equal in length, sides that share a corner — *adjacent sides* — may have different lengths.

- ✔ **Rhombus:** Imagine taking a square and collapsing it as if its corners were hinges. This shape is called a *rhombus*. All four sides are equal in length, and both pairs of opposite sides are parallel.

- ✔ **Parallelogram:** Imagine taking a rectangle and collapsing it as if the corners were hinges. This shape is a *parallelogram* — both pairs of opposite sides are equal in length, and both pairs of opposite sides are parallel.

- ✔ **Trapezoid:** The *trapezoid*'s only important feature is that at least two opposite sides are parallel.

- ✔ **Kite:** A kite is a quadrilateral with two pairs of adjacent sides that are the same length.

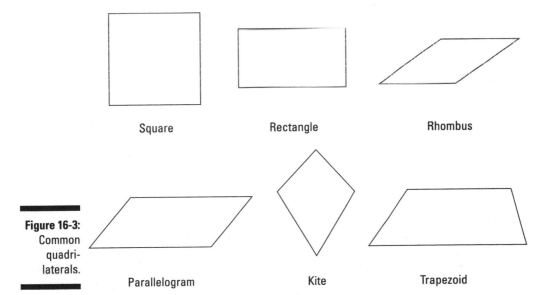

Square   Rectangle   Rhombus

Figure 16-3:
Common
quadri-
laterals.

Parallelogram   Kite   Trapezoid

A quadrilateral can fit into more than one of these categories. For example, every parallelogram (with two sets of parallel sides) is also a trapezoid (with at least one set of parallel sides). Every rectangle and rhombus is also both a parallelogram and a trapezoid. And every square is also all five other types of quadrilaterals. In practice, however, it's common to identify a quadrilateral as descriptively as possible — that is, use the *first* word in the list above that accurately describes it.

Polygons on steroids — larger polygons

A polygon can have any number of sides. Polygons with more than four sides are not as common as triangles and quadrilaterals, but are still worth knowing about. Larger polygons come in two basic varieties: regular and irregular.

A *regular polygon* has equal sides and equal angles. The most common are regular pentagons (5 sides), regular hexagons (6 sides), and regular octagons (8 sides). See Figure 16-4.

Figure 16-4:
A pentagon, a hexagon, and an octagon.

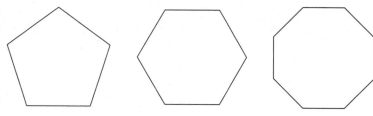

Every other polygon is an *irregular polygon* (see Figure 16-5).

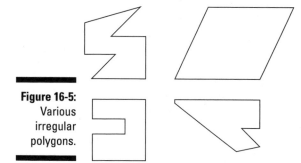

Figure 16-5:
Various irregular polygons.

Taking a Trip to Another Dimension: Solid Geometry

Solid geometry is the study of shapes in *space* — that is, the study of shapes in three dimensions. A *solid* is the spatial (three-dimensional, or 3-D) equivalent of a shape. Every solid has an *inside* and an *outside* separated by the *surface* of the solid. Here, I introduce you to a variety of solids.

The many faces of polyhedrons

A *polyhedron* is the three-dimensional equivalent of a polygon. As you may recall from earlier in the chapter, a polygon is a shape that has only straight sides. Similarly, a polyhedron is a solid that has only straight edges and flat faces (that is, faces that are polygons).

The most common polyhedron is the *cube* (see Figure 16-6). As you can see, a cube has 6 flat faces that are polygons — in this case, all of the faces are square — and 12 straight edges. Additionally, a cube has 8 vertexes (corners). Later in this chapter, I show you how to measure the surface area and volume of a cube.

Figure 16-6:
A typical
cube.

Figure 16-7 shows a few common polyhedrons.

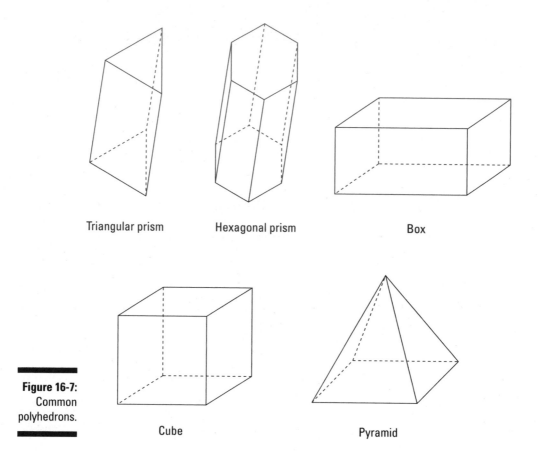

Triangular prism   Hexagonal prism   Box

Figure 16-7:
Common
polyhedrons.

Cube   Pyramid

Later in this chapter, I show you how measure each of these polyhedrons
to determine its volume — that is, the amount of space contained inside its
surface.

One special set of polyhedrons is called the *five regular solids* (see Figure 16-8).
Each regular solid has identical faces that are regular polygons. Notice that
a cube is a type of regular solid. Similarly, the tetrahedron is a pyramid with
four faces that are equilateral triangles.

Tetrahedron   Cube   Octahedron

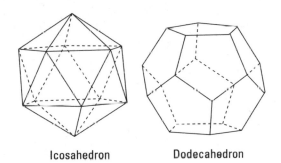

Figure 16-8:
The five
regular
solids.

Icosahedron   Dodecahedron

3-D shapes with curves

Many solids aren't polyhedrons because they contain at least one curved surface. Here are a few of the most common of these types of solids (also see Figure 16-9):

- ✔ **Sphere:** A *sphere* is the solid, or three-dimensional, equivalent of a circle. A ball is a perfect visual aid for a sphere.

- ✔ **Cylinder:** A *cylinder* has a circular base and extends vertically from the plane. A good visual aid for a cylinder is a can of soup.

- ✔ **Cone:** A *cone* is a solid with a round base that extends vertically to a single point. A good visual aid for a cone is an ice cream cone.

Figure 16-9:
Spheres,
cylinders,
and cones.

Sphere   Cylinder   Cone

In the next section, I show you how to measure each of these solids to determine its volume — that is, the amount of space contained inside its surface.

Measuring Shapes: Perimeter, Area, Surface Area, and Volume

In this section, I introduce you to some important formulas for measuring shapes on the plane and solids in space. These formulas use letters to stand for numbers that you can plug in to make specific measurements. Using letters in place of numbers is a feature you'll see more of in Part V, when I discuss algebra.

2-D: Measuring on the flat

Two important skills in geometry — and real life — are finding the perimeter and the area of shapes. A shape's *perimeter* is a measurement of the length of its sides. You use perimeter for measuring the distance around the edges of a room, building, or circular pathway. A shape's *area* is a measurement of how big it is inside. You use area when measuring the size of a wall, a table, or a tire.

For example, in Figure 16-10, I give you the lengths of the sides of each shape.

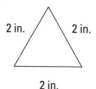

Figure 16-10: Measuring the sides of figures.

2 in.   2 in.   2 in.   1 in.   2 in.   2 in.

2 in.   3 in.   2 in.

When every side of a shape is straight, you can measure its perimeter by adding up the lengths of all of its sides.

Similarly, in Figure 16-11, I give you the area of each shape.

Figure 16-11: The areas of figures.

4 in.²   3 in.²   1.73 in.²

REMEMBER

The area of a shape is always measured in *square units:* square inches (in.^{2}), square feet (ft.^{2}), square miles (mi.^{2}), square kilometers (km^2), and so on — even if you're talking about the area of a circle! (For more on measurements, flip to Chapter 15.)

I cover these types of calculations in this section. (For more information on the names of shapes, refer back to "Closed Encounters: Shaping Up Your Understanding of 2-D Shapes.")

Going 'round in circles

The *center* of a circle is a point that's the same distance from any point on the circle itself. This distance is called the *radius* of the circle, or *r* for short. And any line segment from one point on the circle through the center to another point on the circle is called a *diameter,* or *d* for short. See Figure 16-12.

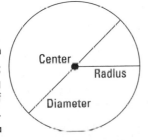

Figure 16-12:
Deciphering
the parts of
a circle.

Center
Radius
Diameter

As you can see, the diameter of any circle is made up of one radius plus another radius — that is, two radii (pronounced *ray*-dee-eye). This concept gives you the following handy formula.

$$d = 2 \cdot r$$

For example, given a circle with a radius of 5 millimeters, you can figure out the diameter as follows:

$$d = 2 \cdot 5 \text{ mm} = 10 \text{ mm}$$

Because the circle is an extra-special shape, its perimeter (the length of its "sides") has an extra-special name: the *circumference* (*C* for short). Early mathematicians went to a lot of trouble figuring out how to measure the circumference of a circle. Here's the formula they hit upon:

$$C = 2 \cdot \pi \cdot r$$

Note: Because $2 \cdot r$ is the same as the diameter, you also can write the formula as $C = \pi \cdot d$.

The symbol π is called *pi* (pronounced *pie*). It's just a number whose approximate value is as follows (the decimal part of pi goes on forever, so you can't get an exact value for pi):

$$\pi \approx 3.14$$

So given a circle with a radius of 5 mm, you can figure out the approximate circumference:

$$C \approx 2 \cdot 3.14 \cdot 5 \text{ mm} = 31.4 \text{ mm}$$

The formula for the area *(A)* of a circle also uses π:

$$A = \pi \cdot r^2$$

Here's how to use this formula to find the approximate area of a circle with a radius of 5 mm:

$$A \approx 3.14 \cdot (5 \text{ mm})^2 = 3.14 \cdot 25 \text{ mm}^2 = 78.5 \text{ mm}^2$$

Measuring triangles

In this section, I discuss how to measure the perimeter and area of all triangles. Then, I show you a special feature of right triangles that allows you to measure them more easily.

Finding the perimeter and area of a triangle

Mathematicians have no special formula for finding the perimeter of a triangle — they just add up the lengths of the sides.

To find the area of a triangle, you need to know the length of one side — the *base* (*b* for short) — and the *height (h)*. Note that the height forms a right angle with the base. Figure 16-13 shows a triangle with a base of 5 cm and a height of 2 cm:

Figure 16-13:
The base
and height
of a triangle.

2 cm

5 cm

Here's the formula for the area of a triangle:

$$A = \frac{1}{2}(b \cdot h)$$

So here's how to figure out the area of a triangle with a base of 5 cm and a height of 2 cm:

$$A = \frac{1}{2}(5 \text{ cm} \cdot 2 \text{ cm}) = \frac{1}{2}(10 \text{ cm}^2) = 5 \text{ cm}^2$$

Lessons from Pythagoras: Finding the third side of a right triangle

The long side of a right triangle (c) is called the *hypotenuse,* and the two short sides $(a$ and $b)$ are called the *legs* (see Figure 16-14). The most important right triangle formula is the *Pythagorean theorem:*

$$a^2 + b^2 = c^2$$

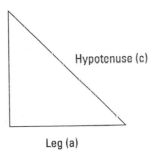

Figure 16-14:
The hypotenuse and legs of a right triangle.

Leg (b)

Hypotenuse (c)

Leg (a)

This formula allows you to find the hypotenuse of a triangle given only the lengths of the legs. For example, suppose the legs of a triangle are 3 and 4 units. Here's how to use the Pythagorean theorem to find the length of the hypotenuse:

$$3^2 + 4^2 = c^2$$
$$9 + 16 = c^2$$
$$25 = c^2$$

So when you multiply c by itself, the result is 25. Therefore,

$$c = 5$$

The length of the hypotenuse is 5 units.

Measuring squares

The letter *s* represents the length of a square's side. For example, if the side of a square is 3 inches, then you say *s* = 3 in. Finding the perimeter *(P)* of a square is simple: Just multiply the length of the side by 4. Here's the formula for the perimeter of a square:

$$P = 4 \cdot s$$

For example, if the length of the side is 3 inches, substitute 3 inches for *s* in the formula:

$$P = 4 \cdot 3 \text{ in.} = 12 \text{ in.}$$

Finding the area of a square is also easy: Just multiply the length of the side by itself — that is, take the *square* of the side. Here are two ways of writing the formula for the area of a square (s^2 is pronounced *s squared*):

$$A = s^2 \quad \text{or} \quad A = s \cdot s$$

For example, if the length of the side is 3 inches, then you get the following:

$$A = (3 \text{ in.})^2 = 3 \text{ in.} \cdot 3 \text{ in.} = 9 \text{ in.}^2$$

Working with rectangles

The long side of a rectangle is called the *length,* or *l* for short. The short side is called the *width,* or *w* for short. For example, in a rectangle whose sides are 5 and 4 feet long, *l* = 5 ft. and *w* = 4 ft.

Because a rectangle has two lengths and two widths, you can use the following formula for the perimeter of a rectangle:

$$P = 2 \cdot (l + w)$$

Calculate the perimeter of a rectangle whose length is 5 yards and whose width is 2 yards as follows:

$$P = 2 \cdot (5 \text{ yd.} + 4 \text{ yd.}) = 2 \cdot 9 \text{ yd.} = 18 \text{ yd.}$$

The formula for the area of a rectangle is:

$$A = l \cdot w$$

So here's how you calculate the area of the same rectangle:

$$A = 5 \text{ yd.} \cdot 4 \text{ yd.} = 20 \text{ yd.}^2$$

Calculating with rhombuses

As with a square, use *s* to represent the length of a rhombus's side. But another key measurement for a rhombus is its height. The *height* of a rhombus (*h* for short) is the shortest distance from one side to the opposite side. In the Figure 16-15, *s* = 4 cm and *h* = 2 cm.

Figure 16-15:
Measuring
a rhombus.

4 cm

2 cm

4 cm

The formula for the perimeter of a rhombus is the same as for a square:

$P = 4 \cdot s$

Here's how you figure out the perimeter of a rhombus whose side is 4 centimeters:

$P = 4 \cdot 4 \text{ cm} = 16 \text{ cm}$

To measure the area of a rhombus, you need both the length of the side and the height. Here's the formula:

$A = s \cdot h$

So here's how you determine the area of a rhombus with a side of 4 cm and a height of 2 cm:

$A = 4 \text{ cm} \cdot 2 \text{ cm} = 8 \text{ cm}^2$

You can read 8 cm² as 8 square centimeters or, less commonly, as 8 centimeters squared.

Measuring parallelograms

The top and bottom sides of a parallelogram are called its *bases* (*b* for short), and the remaining two sides are its *sides* (*s*). And as with rhombuses, another important measurement of a parallelogram is its *height* (*h*), the shortest distance between the bases. So the parallelogram in Figure 16-16 has these measurements: *b* = 6 in., *s* = 3 in., and *h* = 2 in.

Each parallelogram has two equal bases and two equal sides. Therefore, here's the formula for the perimeter of a parallelogram:

$$P = 2 \cdot (b + s)$$

To figure out the perimeter of the parallelogram in this section, just substitute the measurements for the bases and sides:

$$P = 2 \cdot (6 \text{ in.} + 3 \text{ in.}) = 2 \cdot 9 \text{ in.} = 18 \text{ in.}$$

And here's the formula for the area of a parallelogram:

$$A = b \cdot h$$

Here's how you calculate the area of the same parallelogram:

$$A = 6 \text{ in.} \cdot 2 \text{ in.} = 12 \text{ in.}^2$$

Measuring trapezoids

The parallel sides of a trapezoid are called its bases. Because these bases are different lengths, you can call them b_1 and b_2. The *height (h)* of a trapezoid is the shortest distance between the bases. Thus, the trapezoid in Figure 16-17 has these measurements: $b_1 = 2$ in., $b_2 = 3$ in., and $h = 2$ in.

Because a trapezoid can have sides of four different lengths, you really don't have a special formula for finding the perimeter of a trapezoid. Just add up the lengths of its sides and you get your answer.

Here's the formula for the area of a trapezoid:

$$A = \frac{1}{2} \cdot \left(b_1 + b_2\right) \cdot h$$

So, here's how to find the area of the pictured trapezoid:

$$A = \frac{1}{2} \cdot (2 \text{ in.} + 3 \text{ in.}) \cdot 2 \text{ in.}$$
$$= \frac{1}{2} \cdot 5 \text{ in.} \cdot 2 \text{ in.}$$
$$= \frac{1}{2} \cdot 10 \text{ in.}^2 = 5 \text{ in.}^2$$

Note: Because of the associative property (see Chapter 4), I'm allowed to multiply 5 in. · 2 in. before multiplying by ½.

Spacing out: Measuring in three dimensions

In three dimensions, the concepts of perimeter and area have to be tweaked a little.

Recall that in 2-D, the perimeter of a shape is the measurement of its boundary, and the area of a shape is the measurement of what's inside the shape. In 3-D, the boundary of a solid is called its *surface area* and what's inside a solid is called its *volume*.

The *surface area* of a solid is a measurement of the size of its surface, as measured in square units such as square inches (in.^{2}), square feet (ft.^{2}), square meters (m^2), and so forth. The *volume* (V) of a solid is a measurement of the space it occupies, as measured in cubic units such as cubic inches (in.^{3}), cubic feet (ft.^{3}), cubic meters (m^3), and so forth. (For info on measurement, flip to Chapter 15.)

You can find the surface area of a *polyhedron* (solid whose faces are all polygons — see the earlier "The many faces of polyhedrons" section) by adding together the areas of all its faces. The preceding sections give you some area formulas. In most other cases, you don't need to know the formulas for finding the surface area of a solid. (For the names of solids, see the earlier section titled "Taking a Trip to Another Dimension: Solid Geometry.")

Finding the volume of solids, however, is something mathematicians love for you to know. In the next sections, I give you the formulas for finding the volumes of a variety of solids.

Spheres

The *center* of a sphere is a point that's the same distance from any point on the sphere itself. This distance is called the *radius (r)* of the sphere. If you know the radius of a sphere, you can find out its volume using the following formula:

$$V = \frac{4}{3} \cdot \pi \cdot r^3$$

Because this formula includes π, using 3.14 as an approximate value for π gives you an approximation of the volume. For example, here's how to figure out the approximate volume of a ball whose radius is 4 inches:

$$V \approx \frac{4}{3} \cdot 3.14 \cdot (4 \text{ in.})^3$$
$$= \frac{4}{3} \cdot 3.14 \cdot 64 \text{ in.}^3$$
$$\approx 4.19 \cdot 64 \text{ in.}^3$$
$$= 268.16 \text{ in.}^3$$

Cubes

The main measurement of a cube is the length of its side *(s)*. Using this measurement, you can find out the volume of a cube using the following formula:

$$V = s^3$$

So if the side of a cube is 5 meters, here's how you figure out its volume:

$$V = (5 \text{ m})^3 = 5 \text{ m} \cdot 5 \text{ m} \cdot 5 \text{ m} = 125 \text{ m}^3$$

You can read *125 m³* as *125 cubic meters* or, less commonly, as *125 meters cubed*.

Boxes (Rectangular solids)

The three measurements of a box (or rectangular solid) are its length *(l)*, width *(w)*, and height *(h)*. The box pictured in Figure 16-18 has the following measurements: $l = 4$ m, $w = 3$ m, and $h = 2$ m.

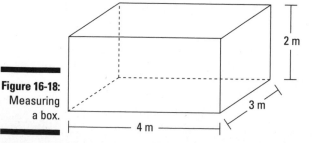

Figure 16-18:
Measuring
a box.

2 m

3 m

4 m

You can find the volume of a box using the following formula:

$$V = l \cdot w \cdot h$$

So here's how to find the volume of the box pictured in this section:

$$V = 4 \text{ m} \cdot 3 \text{ m} \cdot 2 \text{ m} = 24 \text{ m}^3$$

Prisms

Finding the volume of a prism (see prisms in Figure 16-7) is easy if you have two measurements. One measurement is the *height (h)* of the prism. The second is the *area of the base (A_b)*. The *base* is the polygon that extends vertically from the plane. (In "2-D: Measuring on the flat" earlier, I show you how to find the area of a variety of shapes.)

Here's the formula for finding the volume of a prism:

$$V = A_b \cdot h$$

For example, suppose a prism has a base with an area of 5 square centimeters and a height of 3 centimeters. Here's how you find its volume:

$$V = 5 \text{ cm}^2 \cdot 3 \text{ cm} = 15 \text{ cm}^3$$

Notice that the units of measurements (cm^2 and cm) are also multiplied, giving you a result of cm^3.

Cylinders

You find the volume of cylinders the same way you find the area of prisms — by multiplying the area of the base (A_b) by the cylinder's height *(h)*:

$$V = A_b \cdot h$$

Suppose you want to find the volume of a cylindrical can whose height is 4 inches and whose base is a circle with a radius of 2 inches. First, find the area of the base by using the formula for the area of a circle:

$$A_b = \pi \cdot r^2$$
$$\approx 3.14 \cdot (2 \text{ in.})^2$$
$$= 3.14 \cdot 4 \text{ in.}^2$$
$$= 12.56 \text{ in.}^2$$

This area is *approximate* because I use 3.14 as an approximate value for π. (**Note:** In the preceding problem, I use equal signs when a value is equal to whatever comes right before it and approximately-equal-to signs ($\approx$) when I round.)

Now use this area to find the volume of the cylinder:

$$V \approx 12.56 \text{ in.}^2 \cdot 4 \text{ in.} = 50.24 \text{ in.}^3$$

Notice how multiplying square inches (in.^{2}) by inches gives a result in cubic inches (in.^{3}).

Pyramids and cones

The two key measurements for pyramids and cones are the same as those for prisms and cylinders (see the previous sections): the *height (h)* and the area of the base *(A_b)*. Here's the formula for the volume of a pyramid or a cone:

$$V = \frac{1}{3}\left(A_b \cdot h\right)$$

For example, suppose you want to find the volume of an ice cream cone whose height is 4 inches and whose base area is 3 square inches. Here's how you do it:

$$V = \frac{1}{3}\left(3 \text{ in.}^2 \cdot 4 \text{ in.}\right)$$
$$= \frac{1}{3}\left(12 \text{ in.}^3\right)$$
$$= 4 \text{ in.}^3$$

Similarly, suppose you want to find the volume of a pyramid in Egypt whose height is 60 meters with a square base whose sides are each 50 meters. First, find the area of the base using the formula for the area of a square from "2-D: Measuring on the flat," earlier in this chapter:

$$A_b = s^2 = (50 \text{ m})^2 = 2{,}500 \text{ m}^2$$

Now use this area to find the volume of the pyramid:

$$V = \frac{1}{3}\left(2{,}500 \text{ m}^2 \cdot 60 \text{ m}\right)$$
$$= \frac{1}{3}\left(150{,}000 \text{ m}^3\right)$$
$$= 50{,}000 \text{ m}^3$$

Chapter 17

Seeing Is Believing: Graphing as a Visual Tool

A *graph* is a visual tool for organizing and presenting information about numbers. Most students find graphs relatively easy because they provide a picture to work with rather than just a bunch of numbers. Their simplicity makes graphs show up in newspapers, magazines, business reports, and anywhere clear visual communication is important.

In this chapter, I introduce you to three common styles of graphs: the bar graph, the pie chart, and the line graph. I show you how to read each of these styles of graphs to obtain information. I also show you how to answer the types of questions people may ask to check your understanding.

I spend the rest of the chapter focusing on the most important type of mathematical graph: the Cartesian coordinate system. This system is so common that when math folks talk about a graph, they're usually talking about this type. I show you the various parts of the graph, and then I show you how to plot points and lines. At the end of the chapter, you see how you can solve math problems using a graph.

Looking at Three Important Graph Styles

In this section, I show you how to read and understand three styles of graphs: the bar graph, the pie chart, and the line graph. They aren't the only types of graphs, but they're very common, and understanding them can give you a leg up on reading other types of graphs when you see them.

Each of these graph styles has a specific function:

- ✔ The bar graph is best for representing numbers that are independent of each other.
- ✔ The pie chart allows you to show how a whole is cut up into parts.
- ✔ The line graph gives you a sense of how numbers change over time.

Bar graph

A *bar graph* gives you an easy way to compare numbers or values. For example, Figure 17-1 shows a bar graph comparing the performance of five trainers at a fitness center.

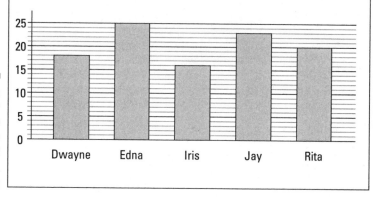

Figure 17-1: The number of new clients recorded this quarter.

As you can see from the caption, the graph shows how many new clients each trainer has enrolled this quarter. The advantage of such a graph is that you can see at a glance, for example, that Edna has the most new clients and Iris has the fewest. The bar graph is a good way to represent numbers that are independent of each other. For example, if Iris gets another new client, it doesn't necessarily affect any other trainer's performance.

Reading a bar graph is easy after you get used to it. Here are a few types of questions someone could ask about the bar graph in Figure 17-1:

- **Individual values:** *How many new clients does Jay have?* Find the bar representing Jay's clients and notice that he has 23 new clients.

- **Differences in value:** *How many more clients does Rita have compared with Dwayne's?* Notice that Rita has 20 new clients and Dwayne has 18, so she has 2 more than he does.

- **Totals:** *Together, how many clients do the three women have?* Notice that the three women — Edna, Iris, and Rita — have 25, 16, and 20 new clients, respectively, so they have 61 new clients altogether.

Pie chart

A *pie chart,* which looks like a divided circle, shows you how a whole object is cut up into parts. Pie charts are most often used to represent percentages. For example, Figure 17-2 is a pie chart representing Eileen's monthly expenses.

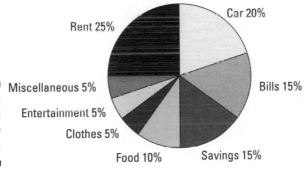

Figure 17-2:
Eileen's
monthly
expenses.

You can tell at a glance that Eileen's largest expense is rent and that her second largest is her car. Unlike the bar graph, the pie chart shows numbers that are dependent upon each other. For example, if Eileen's rent increases to 30% of her monthly income, she'll have to decrease her spending in at least one other area.

Here are a few typical questions you may be asked about a pie chart:

- **Individual percentages:** *What percentage of her monthly expenses does Eileen spend on food?* Find the slice that represents what Eileen spends on food, and notice that she spends 10% of her income there.

- ✔ **Differences in percentages:** *What percentage more does she spend on her car than on entertainment?* Eileen spends 20% on her car but only 5% on entertainment, so the difference between these percentages is 15%.

- ✔ **How much a percent represents in terms of dollars:** *If Eileen brings home $2,000 per month, how much does she put away in savings each month?* First notice that Eileen puts 15% every month into savings. So you need to figure out 15% of $2,000. Using your skills from Chapter 12, solve this problem by turning 15% into a decimal and multiplying:

  $$0.15 \cdot 2,000 = 300$$

  So Eileen saves $300 every month.

Line graph

The most common use of a *line graph* is to plot how numbers change over time. For example, Figure 17-3 is a line graph showing last year's sales figures for Tami's Interiors.

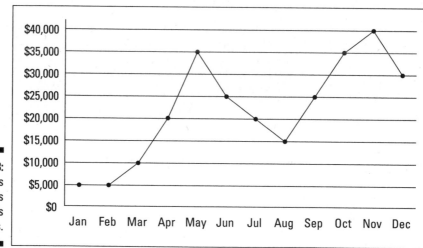

Figure 17-3:
Gross
receipts
for Tami's
Interiors.

The line graph shows a progression in time. At a glance, you can tell that Tami's business tended to rise strongly at the beginning of the year, drop off during the summer, rise again in the fall, and then drop off again in December.

Here are a few typical questions that you may be asked to show that you know how to read a line graph:

- **High or low points and timing:** *In what month did Tami bring in the most revenue, and how much did she bring in?* Notice that the highest point on the graph is in November, when Tami's revenue reached $40,000.

- **Total over a period of time:** *How much did she bring in altogether the last quarter of the year?* A quarter of a year is three months, so the last quarter is the last three months of the year. Tami brought in $35,000 in October, $40,000 in November, and $30,000 in December, so her total receipts for the last quarter add up to $105,000.

- **Greatest change:** *In what month did the business show the greatest gain in revenue as compared with the previous month?* You want to find the line segment on the graph that has the steepest upward slope. This change occurs between April and May, where Tami's revenue increased by $15,000, so her business showed the greatest gain in May.

Cartesian Coordinates

When math folks talk about using a graph, they're usually referring to a *Cartesian graph* (also called the Cartesian coordinate system), as shown in Figure 17-4. In Chapter 25, I tell you why I believe that this graph is one of the ten most important mathematical inventions of all time. You see a lot of this graph when you study algebra, so getting familiar with it now is a good idea.

Figure 17-4:
A Cartesian graph includes horizontal and vertical axes, which cross at the origin (0, 0).

A Cartesian graph is really just two number lines that cross at 0. These number lines are called the *horizontal axis* (also called the *x-axis*) and the *vertical axis* (also called the *y-axis*). The place where these two axes (plural of axis) cross is called the *origin*.

Plotting points on a Cartesian graph

Plotting a point (finding and marking its location) on a graph isn't much harder than finding a point on a number line, because a graph is just two number lines put together. (Flip to Chapter 1 for more on using the number line.)

Every point on a Cartesian graph is represented by two numbers in parentheses, separated by a comma, called a set of *coordinates*. To plot any point, start at the origin, where the two axes cross. The first number tells you how far to go to the right (if positive) or left (if negative) along the horizontal axis. The second number tells you how far to go up (if positive) or down (if negative) along the vertical axis.

For example, here are the coordinates of four points called *A, B, C,* and *D:*

$$A = (2, 3) \qquad B = (-4, 1) \qquad C = (0, -5) \qquad D = (6, 0)$$

Figure 17-5 depicts a graph with these four points plotted. Start at the origin, (0, 0). To plot point *A,* count 2 spaces to the right and 3 spaces up. To plot point *B,* count 4 spaces to the left (the negative direction) and then 1 space up. To plot point *C,* count 0 spaces left or right and then count 5 spaces down (the negative direction). And to plot point *D,* count 6 spaces to the right and then 0 spaces up or down.

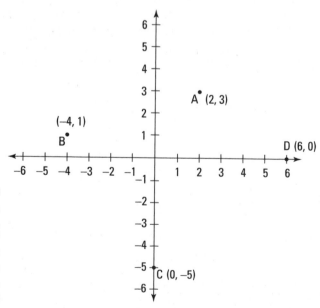

Figure 17-5:
Points *A, B, C,* and *D* plotted on a Cartesian graph.

Drawing lines on a Cartesian graph

After you understand how to plot points on a graph (see the preceding section), you can begin to plot lines and use them to show mathematical relationships.

The examples in this section focus on the number of dollars that two people, Xenia and Yanni, are carrying. The horizontal axis represents Xenia's money, and the vertical axis represents Yanni's. For example, suppose you want to draw a line representing this statement:

Xenia has $1 more than Yanni.

To do this, make a chart as follows:

Xenia	1	2	3	4	5
Yanni					

Now fill in each column of the chart assuming that Xenia has that number of dollars. For example, if Xenia has $1, then Yanni has $0. And if Xenia has $2, then Yanni has $1. Continue until your chart looks like this:

Xenia	1	2	3	4	5
Yanni	0	1	2	3	4

Now you have five pairs of points that you can plot on your graph as (Xenia, Yanni): (1, 0), (2, 1), (3, 2), (4, 3), and (5, 4). Next, draw a straight line through these points, as shown in Figure 17-6.

This line on the graph represents every possible pair of amounts for Xenia and Yanni. For example, notice how the point (6, 5) is on the line. This point represents the possibility that Xenia has $6 and Yanni has $5.

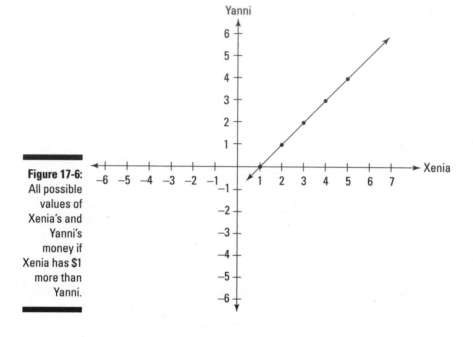

Figure 17-6:
All possible
values of
Xenia's and
Yanni's
money if
Xenia has $1
more than
Yanni.

Here's a slightly more complicated example:

Yanni has $3 more than twice the amount that Xenia has.

Again, start by making the usual chart:

Xenia	1	2	3	4	5
Yanni					

You can fill in this chart by supposing that Xenia has a certain amount of money and then figuring out how much money Yanni would have in that case. For example, if Xenia has $1, then twice that amount is $2, so Yanni has $3 more than that, or $5. And if Xenia has $2, then twice that amount is $4, so Yanni has $3 more, or $7. Continue in that way to fill in the chart as follows:

Xenia	1	2	3	4	5
Yanni	5	7	9	11	13

Now plot these five points on the graph and draw a line through them, as in Figure 17-7.

As in the other examples, this graph represents all possible values that Xenia and Yanni could have. For example, if Xenia has $7, Yanni has $17.

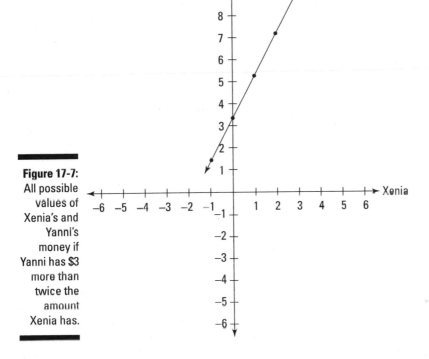

Figure 17-7:
All possible
values of
Xenia's and
Yanni's
money if
Yanni has $3
more than
twice the
amount
Xenia has.

Solving problems with a Cartesian graph

After you understand how to plot points and draw lines, you can use a graph to solve certain types of math problems. When you draw two lines that represent different parts of a word problem, then the point at which the lines intersect (where they cross) is your answer.

Here's an example:

> Jacob is exactly 5 years younger than Marnie, and together their ages add up to 15. How old are they?

To solve this problem, first make a chart to show that Jacob is 5 years younger than Marnie:

Jacob	1	2	3	4	5
Marnie	6	7	8	9	10

Then make another chart to show that together, the two children's ages add up to 15:

Jacob	1	2	3	4	5
Marnie	14	13	12	11	10

Finally, plot both lines on a graph (see Figure 17-8) where the horizontal axis represents Jacob's age and the vertical axis represents Marnie's age. Notice that the two lines cross at the point where Jacob is 5 and Marnie is 10 years old, so these are the two children's ages.

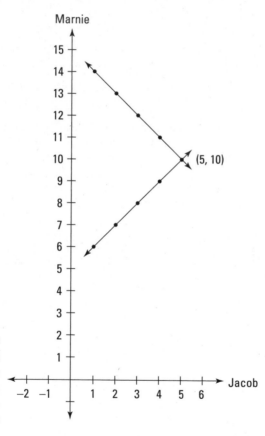

Figure 17-8:
Both lines plotted on a graph.

Chapter 18

Solving Geometry and Measurement Word Problems

In This Chapter

▶ Solving measurement problems using conversion chains

▶ Using a picture to solve geometry problems

*I*n this chapter, I focus on two important types of word problems: measurement problems and geometry problems. In a word problem involving measurement, you're often asked to perform a conversion from one type of unit to another. Sometimes you don't have a conversion equation to solve this type of problem directly, so you need to set up a *conversion chain,* which I discuss in detail in the chapter.

Another common type of word problem requires the geometric formulas that I provide in Chapter 16. Sometimes, a geometry word problem gives you a picture to work with. In other cases, you have to draw the picture yourself by reading the problem carefully. Here I give you practice doing both types of problems.

The Chain Gang: Solving Measurement Problems with Conversion Chains

In Chapter 15, I give you a set of basic conversion equations for converting units of measurement. I also show you how to turn these equations into conversion factors — fractions that you can use to convert units. This information is useful as far as it goes, but you may not always have an equation for the exact conversion that you want to perform. For example, how do you convert years to seconds?

For more-complex conversion problems, a good tool is the conversion chain. A *conversion chain* links together a sequence of unit conversions.

Setting up a short chain

Here's a problem that shows you how to set up a short conversion chain to make a conversion you won't find a specific equation for:

> Vendors at the Fragola County Strawberry Festival sold 7 tons of strawberries in a single weekend. How many 1-ounce servings of strawberries is that?

You don't have an equation to convert tons directly to ounces. But you do have one to convert tons to pounds and another to convert pounds to ounces. You can use these equations to build a bridge from one unit to another:

tons → pounds → ounces

So here are the two equations that you'll want to use:

1 ton = 2,000 lbs.

1 lb. = 16 oz.

To convert tons to pounds, note that these fractions equal 1, because the numerator (top number) equals the denominator (bottom number):

$$\frac{1 \text{ ton}}{2{,}000 \text{ lbs.}} \quad \text{or} \quad \frac{2{,}000 \text{ lbs.}}{1 \text{ ton}}$$

To convert pounds to ounces, note that these fractions equal 1:

$$\frac{1 \text{ lb.}}{16 \text{ oz.}} \quad \text{or} \quad \frac{16 \text{ oz.}}{1 \text{ lb.}}$$

You could do this conversion in two steps. But when you know the basic idea, you set up a conversion chain instead. To help make this idea clear, take a look at how to get from tons to ounces:

tons → pounds → ounces

So here's how to set up a conversion chain to turn 7 tons into pounds and then into ounces. Because you already have tons on top, you want the tons-and-pounds fraction that puts *ton* on the bottom. And because that fraction puts *pounds* on the top, use the pounds-and-ounces fraction that puts *pound* on the bottom:

$$\frac{7 \text{ tons}}{1} \cdot \frac{2{,}000 \text{ lbs.}}{1 \text{ ton}} \cdot \frac{16 \text{ oz.}}{1 \text{ lb.}}$$

The net effect here is to take the expression *7 tons* and multiply it twice by 1, which doesn't change the value of the expression. But now, you can cancel out all units of measurement that appear in the numerator of one fraction and the denominator of another:

$$= \frac{7 \text{ tons}}{1} \cdot \frac{2,000 \text{ lbs.}}{1 \text{ ton}} \cdot \frac{16 \text{ oz.}}{1 \text{ lb.}}$$

If any units don't cancel out properly, you probably made a mistake when you set up the chain. Flip the numerator and denominator of one or more of the fractions until the units cancel out the way you want them to.

Now you can simplify the expression:

$$= 7 \cdot 2,000 \cdot 16 \text{ oz.}$$

A conversion chain doesn't change the *value* of the expression, just the units of measurement.

Working with more links

After you understand the basic idea of a conversion chain, you can make a chain as long as you like to solve longer problems easily. Here's another example of a problem that uses a time-related conversion chain:

> Jane is exactly 12 years old today. You forgot to get her a present, but you decide that offering her your mathematical skills is the greatest gift of all — you'll recalculate how old she is. Assuming that a year has exactly 365 days, how many seconds old is she?

Here are the conversion equations you have to work with:

> 1 year = 365 days
>
> 1 day = 24 hours
>
> 1 hour = 60 minutes
>
> 1 minute = 60 seconds

To solve this problem, you need to build a bridge from years to seconds as follows:

> years → days → hours → minutes → seconds

So set up a long conversion chain as follows:

$$\frac{12 \text{ years}}{1} \cdot \frac{365 \text{ days}}{1 \text{ year}} \cdot \frac{24 \text{ hrs.}}{1 \text{ day}} \cdot \frac{60 \text{ min.}}{1 \text{ hr.}} \cdot \frac{60 \text{ sec.}}{1 \text{ min.}}$$

Cancel out all units that appear in both a numerator and a denominator:

$$= \frac{12 \, \cancel{\text{years}}}{1} \cdot \frac{365 \, \cancel{\text{days}}}{1 \, \cancel{\text{year}}} \cdot \frac{24 \, \cancel{\text{hrs.}}}{1 \, \cancel{\text{day}}} \cdot \frac{60 \, \cancel{\text{min.}}}{1 \, \cancel{\text{hr.}}} \cdot \frac{60 \text{ sec.}}{1 \, \cancel{\text{min.}}}$$

As you cancel out units, notice that there is a *diagonal* pattern: the numerator (top number) of one fraction cancels with the denominator (bottom number) of the next, and so on.

When the smoke clears, here's what's left:

$$= 12 \cdot 365 \cdot 24 \cdot 60 \cdot 60 \text{ sec.}$$

This problem requires a bit of multiplication, but the work is no longer confusing:

$$= 378{,}432{,}000 \text{ sec.}$$

The conversion chain from 12 years to 378,432,000 seconds doesn't change the value of the expression, just the unit of measurement.

Pulling equations out of the text

In some word problems, the problem itself gives you a couple of the conversion equations necessary for solving. Take this problem for example:

A furlong is ⅛ of a mile, and a fathom is 2 yards. If I rode my horse 24 furlongs today, how many fathoms did I ride?

This problem gives you two new conversion equations to work with:

1 furlong = ⅛ mile

1 fathom = 2 yards

It's helpful to remove fractions from the equations before you begin, so here's a more useful version of the first equation:

8 furlongs = 1 mile

You also want to remember two other conversions:

1 mile = 5,280 feet

3 feet = 1 yard

Next, build a bridge from furlongs to miles using the conversions that are available from these equations:

furlongs → miles → feet → yards → fathoms

Now you can form your conversion chain. Every unit you want to cancel has to appear once in the numerator and once in the denominator:

$$\frac{24 \text{ furlongs}}{1} \cdot \frac{1 \text{ mile}}{8 \text{ furlongs}} \cdot \frac{5{,}280 \text{ feet}}{1 \text{ mile}} \cdot \frac{1 \text{ yard}}{3 \text{ feet}} \cdot \frac{1 \text{ fathom}}{2 \text{ yards}}$$

Next, you can cancel out all the units except for fathoms:

$$= \frac{24 \, \cancel{\text{furlongs}}}{1} \cdot \frac{1 \text{ mile}}{8 \, \cancel{\text{furlongs}}} \cdot \frac{5{,}280 \, \cancel{\text{feet}}}{1 \, \cancel{\text{mile}}} \cdot \frac{1 \, \cancel{\text{yard}}}{3 \, \cancel{\text{feet}}} \cdot \frac{1 \text{ fathom}}{2 \, \cancel{\text{yards}}}$$

Another way you can make this problem a little easier is to notice that the number 24 is in the numerator and 3 and 8 are in the denominator. Of course, $3 \cdot 8 = 24$, so you can cancel out all three of these numbers:

$$= \frac{\cancel{24 \text{ furlongs}}}{1} \cdot \frac{1 \, \cancel{\text{mile}}}{\cancel{8 \text{ furlongs}}} \cdot \frac{5{,}280 \, \cancel{\text{feet}}}{1 \, \cancel{\text{mile}}} \cdot \frac{1 \, \cancel{\text{yard}}}{\cancel{3 \text{ feet}}} \cdot \frac{1 \text{ fathom}}{2 \, \cancel{\text{yards}}}$$

At this point, the expression has only two numbers left aside from the 1s, and the fraction's easy to simplify:

$$= \frac{5{,}280}{2} \text{ fathoms} - 2{,}640 \text{ fathoms}$$

As always, the conversion chain from 24 furlongs to 2,640 fathoms doesn't change the value of the expression, just the units of measurement.

Rounding off: Going for the short answer

Sometimes real-life measurements just aren't that accurate. After all, If you measure the length of a football field with your trusty ruler, you're bound to be off an inch or two (or more). When you perform calculations with such measurements, finding the answer to a bunch of decimal places doesn't make sense because the answer's already approximate. Instead, you want to round off your answer to the numbers that are probably correct. Here's a problem that asks you to do just that:

> Heather weighed her new pet hamster, Binky, and found that he weighs 4 ounces. How many grams does Binky weigh, to the nearest whole gram?

This problem requires you to convert from English to metric units, so you need this conversion equation:

$$1 \text{ kilogram} \approx 2.20 \text{ pounds}$$

Notice that this conversion equation includes only kilograms and pounds, but the problem includes ounces and grams. So to convert from ounces to pounds and from kilograms to grams, here are some equations to help build a bridge between ounces and grams:

$$1 \text{ pound} = 16 \text{ ounces}$$

$$1 \text{ kilogram} = 1{,}000 \text{ grams}$$

Your chain will perform the following conversions:

$$\text{ounces} \to \text{pounds} \to \text{kilograms} \to \text{grams}$$

So set up your expression as follows:

$$\frac{4 \text{ oz.}}{1} \cdot \frac{1 \text{ lb.}}{16 \text{ oz.}} \cdot \frac{1 \text{ kg}}{2.2 \text{ lb.}} \cdot \frac{1{,}000 \text{ g}}{1 \text{ kg}}$$

As always, after you set up the expression, you can cancel out every unit except for the one you're converting to:

$$= \frac{4 \ \cancel{\text{oz.}}}{1} \cdot \frac{1 \ \cancel{\text{lb.}}}{16 \ \cancel{\text{oz.}}} \cdot \frac{1 \ \cancel{\text{kg}}}{2.2 \ \cancel{\text{lb.}}} \cdot \frac{1{,}000 \text{ g}}{1 \ \cancel{\text{kg}}}$$

 When you're multiplying a string of fractions, you can make one fraction out of all the numbers. The numbers that were originally in the numerators of fractions remain in the numerator. Similarly, the numbers that were in the denominators remain in the denominator. Then just put a multiplication sign between each pair of numbers.

$$= \frac{4 \cdot 1{,}000}{16 \cdot 2.2} \text{ g}$$

At this point, you can begin calculating. But to save some effort, I recommend canceling out common factors. In this case, you cancel out a 4 in the numerator and denominator, changing the 16 in the denominator to a 4:

$$= \frac{4 \cdot 1{,}000}{4\overline{16} \cdot 2.2} \text{ g}$$

Now, you can cancel out another 4 in the numerator and denominator, changing the 1,000 in the numerator to 250:

$$= \frac{4 \cdot \overline{1{,}000}^{250}}{4\,\overline{16} \cdot 2.2} \text{ g}$$

At this point, here's what's left:

$$= \frac{250}{2.2} \text{ g}$$

Divide 250 by 2.2 to get your answer:

$$\approx 113.6 \text{ g}$$

Notice that I took the division out to one decimal place. Because the number after the decimal point is 6, I need to round my answer up to the next highest gram. (See Chapter 11 for more about rounding decimals.)

So to the nearest gram, Binky weighs 114 grams. As usual, the conversion chain doesn't change the value of the expression, just the unit of measurement.

Solving Geometry Word Problems

Some geometry word problems present you with a picture. In other cases, you have to draw a picture yourself. Sketching figures is always a good idea because it can usually give you an idea of how to proceed. The following sections present you with both types of problems. (To solve these word problems, you need some of the geometry formulas I discuss in Chapter 16.)

Working from words and images

Sometimes, you have to interpret a picture to solve a word problem. Read the problem carefully, recognize shapes in the drawing, pay attention to labels, and use whatever formulas you have to help you answer the question. In this problem, you get to work with a picture.

Mr. Dennis is a farmer with two teenage sons. He gave them a rectangular piece of land with a creek running through it diagonally, as shown in Figure 18-1. The elder boy took the larger area and the younger boy took the smaller. What is the area of each boy's land in square feet?

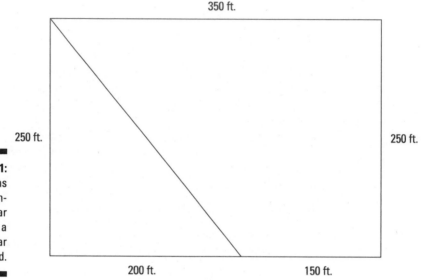

350 ft.

250 ft.   250 ft.

200 ft.   150 ft.

Figure 18-1:
Two sons
get non-
rectangular
portions of a
rectangular
field.

To find the area of the smaller, triangular plot, use the formula for the area of a triangle, where A is the area, b is the base, and h is the height:

$$A = \frac{1}{2}(b \cdot h)$$

The whole piece of land is a rectangle, so you know that the corner the triangle shares with the rectangle is a right angle. Therefore, you know that the sides labeled 200 feet and 250 feet are the base and height. Find the area of this plot by plugging the base and height into the formula:

$$A = \frac{200 \text{ feet} \cdot 250 \text{ feet}}{2}$$

To make this calculation a little easier, notice that you can cancel a factor of 2 from the numerator and denominator:

$$A = \frac{\overset{100}{\cancel{200}} \text{ feet} \cdot 250 \text{ feet}}{\cancel{2}} = 25,000 \text{ square feet}$$

The shape of the remaining area is a trapezoid. You can find its area by using the formula for a trapezoid, but there's an easier way. Because you know the area of the triangular plot, you can use this word equation to find the area of the trapezoid:

Area of trapezoid = area of whole plot − area of triangle

To find the area of the whole plot, remember the formula for the area of a rectangle. Plug its length and width into the formula:

A = length · width

A = 350 ft. · 250 ft.

A = 87,500 square ft.^{2}

Now just substitute the numbers that you know into the word equation you set up:

Area of trapezoid = 87,500 square feet − 25,000 square feet

= 62,500 square feet

So the area of the elder boy's land is 62,500 square feet, and the area of the younger boy's land is 25,000 square feet.

Breaking out those sketching skills

Geometry word problems may not make much sense until you draw some pictures. Here's an example of a geometry problem without a picture provided:

In Elmwood Park, the flagpole is due south of the swing set and exactly 20 meters due west of the treehouse. If the area of the triangle made by the flagpole, the swing set, and the treehouse is 150 square meters, what is the distance from the swing set to the treehouse?

This problem is bound to be confusing until you draw a picture of what it's telling you. Start with the first sentence, depicted in Figure 18-2. You don't need to draw the ladder up to the treehouse, some kids fighting over the swings, and a flag with the correct number of stars and stripes waving at the top of the pole — labels do just fine. As you can see, I've drawn a right triangle whose corners are the swing set (S), the flagpole (F), and the treehouse (T). I've also labeled the distance from the flagpole to the treehouse as 20 meters.

Figure 18-2:
A labeled
sketch
shows the
important
information
in a word
problem.

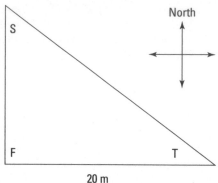

The next sentence tells you the area of this triangle:

$$A = 150 \text{ m}^2$$

Now you're out of information, so you need to remember anything you can from geometry. Because you know the area of the triangle, you may find the formula for the area of a triangle helpful:

$$A = \frac{1}{2}(b \cdot h)$$

Here, b is the base and h is the height. In this case, you have a right triangle, so the base is the distance from F to T and the height is the distance from S to F. So you already know the area of the triangle, and you also know the length of the base. Fill in the equation:

$$150 = \frac{1}{2}(20 \cdot h)$$

You can now solve this equation for h. Start by simplifying:

$$150 = 10 \cdot h$$

You can turn this into a division problem using inverse operations, as I show you in Chapter 4:

$$150 \div 10 = h$$
$$15 = h$$

Now you know that the height of the triangle is 15 m, so you can add this information to your picture, as shown in Figure 18-3.

Figure 18-3:
Update the
labels in
your sketch
as you work
through the
problem.

15 m

20 m

To solve the problem, though, you still need to find out the distance from S to T. Because this is a right triangle, you can use the Pythagorean theorem to figure out the distance:

$$a^2 + b^2 = c^2$$

Remember that a and b are the lengths of the short sides, and c is the length of the longest side, called the hypotenuse. (See Chapter 16 for more on the Pythagorean theorem.) You can substitute numbers into this formula as follows:

$$15^2 + 20^2 = c^2$$

Begin solving for c by evaluating the left side of the equation according to the order of operations (see Chapter 5 for details). Start with the two exponents and then move on to addition:

$$225 + 400 = c^2$$
$$625 = c^2$$

At this point, remember that c^2 means $c \cdot c$:

$$625 = c \cdot c$$

So c is a number that, when multiplied by itself, results in the number 625. It's greater than 20 (because one of the shorter sides of the triangle is 20), but less than 30 (because $30 \cdot 30 = 900$). A bit of trial and error gives you the following:

$$625 = 25 \cdot 25$$

So the distance from the swing set to the treehouse is 25 meters.

Together at Last: Putting Geometry and Measurements in One Problem

Word problems often ask you to use a variety of skills. This is especially popular (among teachers, not students!) on a final exam, when you're being tested on everything that a class covered for a whole semester. So here's an example that pulls together everything from this chapter:

> A circular fountain that's 32 feet across is surrounded by a circular path that's 120 feet in length around at its outside edge. What is the width of the path in inches to the nearest inch? (Use 3.1 as an approximation of π.)

First, draw a picture of what the problem is asking, like the one shown in Figure 18-4.

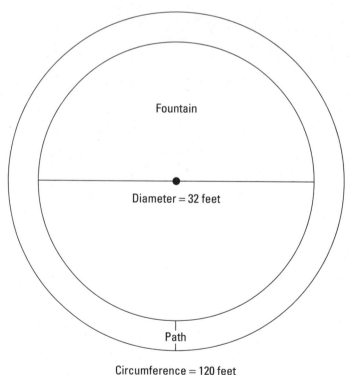

Figure 18-4:
Measuring
a path
around a
fountain.

As you can see, I drew a circle for the fountain and labeled its diameter as 32 feet. The outside edge of the path around the fountain is also a circle, and its circumference is 120 feet. The problem is asking for the width of the path. The picture shows you that the path's width is the distance from the inner circle to the outer circle.

The *radius* of a circle is the distance from its center to the circle itself (see Chapter 16). So if you know the radius of each circle, you can find the width of the path by subtracting:

Width of path = radius of outer circle – radius of inner circle

This word equation is the key to the problem. Look at the diagram and make sure you understand it before continuing.

You already know that the diameter of the inner circle is 32 feet, so you can find its radius using this formula from Chapter 16, where *d* is the diameter and *r* is the radius:

$d = 2 \cdot r$

Plugging in the diameter of the inner circle, you get the following:

32 ft. = 2 · radius of inner circle

You can probably solve this problem in your head:

Radius of inner circle = 16 ft.

You also know the circumference of the outer circle, so you can find its radius using this formula from Chapter 16, where *C* is the circumference and *r* is the radius:

$C = 2 \cdot \pi \cdot r$

Plugging in the circumference and 3.1 for π gives you

120 ft. = 2 · 3.1 · radius of outer circle

This equation can be simplified a little as follows:

120 ft. = 6.2 · radius of outer circle

Now, you can use inverses to rearrange this multiplication problem into a division problem (see Chapter 4 for more on how to do this):

120 ft. ÷ 6.2 = radius of outer circle

Rounding off this division to the nearest tenth gives you this answer:

Radius of outer circle ≈ 19.4 ft.

Now that you have the radius of both circles, you can substitute into the word equation that you started with:

Width of path ≈ 19.4 ft. − 16 ft. = 3.4 ft.

The problem asks for the answer to the nearest inch. Here's a conversion equation:

1 ft. = 12 in.

You can set up an expression as follows:

$$\frac{3.4 \text{ ft.}}{1} \cdot \frac{12 \text{ in.}}{1 \text{ ft.}}$$

Now, you can cancel out *feet,* because this unit appears in both the numerator and denominator:

$$= \frac{3.4 \text{ ft.}}{1} \cdot \frac{12 \text{ in.}}{1 \text{ ft.}}$$

This expression now simplifies to the following:

$$= 3.4 \cdot 12 \text{ in.} = 40.8 \text{ in.}$$

So the width of the path is about 41 inches. This problem is a good one to try on your own. Copy it down and see whether you can work through it from beginning to end.

Chapter 19

Figuring Your Chances: Statistics and Probability

- -

In This Chapter

▶ Knowing how statistics works with both qualitative and quantitative data

▶ Finding out how to calculate a percentage and the mode of a sample

▶ Calculating the mean and median

▶ Finding the probability of an event

- -

Statistics and probability are two of the most important and widely used applications of math. They're applicable to virtually every aspect of the real world — business, biology, city planning, politics, meteorology, and many more areas of study. Even physics, once thought to be devoid of uncertainty, now relies on probability.

In this chapter, I give you a basic understanding of these two mathematical ideas. First, I introduce you to statistics and the important distinction between qualitative and quantitative data. I show you how to work with both types of data to find meaningful answers. Then I give you the basics of probability. I show you how the probability that an event will occur is always a fraction from 0 to 1. After that, I demonstrate how to build this fraction by counting both favorable outcomes and possible outcomes. Finally, I put these ideas to work by showing you how to calculate the probability of tossing coins and rolling dice.

Gathering Data Mathematically: Basic Statistics

Statistics is the science of gathering and drawing conclusions from *data,* which is information that's measured objectively in an unbiased, reproducible way.

An individual *statistic* is a conclusion drawn from this data. Here are some examples:

- ✔ The average working person drinks 3.7 cups of coffee every day.

- ✔ Only 52% of students who enter law school actually graduate.

- ✔ The cat is the most popular pet in the United States.

- ✔ In the last year, the cost of a high-definition TV dropped by an average of $575.

Statisticians do their work by identifying a population that they'd like to study: working people, law students, pet owners, buyers of electronics, whoever. Because most populations are far too large to work with, a statistician collects data from a smaller, randomly selected sample of this population. Much of statistics concerns itself with gathering data that's reliable and accurate. You can read all about this idea in *Statistics For Dummies* (Wiley).

In this section, I give you a short introduction to the more mathematical aspects of statistics.

Understanding differences between qualitative and quantitative data

Data — the information used in statistics — can be either qualitative or quantitative. *Qualitative data* divides a data set (the pool of data that you've gathered) into discrete chunks based on a specific attribute. For example, in a class of students, qualitative data can include

- ✔ Each child's gender

- ✔ His or her favorite color

- ✔ Whether he or she owns at least one pet

- ✔ How he or she gets to and from school

You can identify qualitative data by noticing that it links an *attribute* — that is, a quality — to each member of the data set. For example, four attributes of Emma are that she's female, her favorite color is green, she owns a dog, and she walks to school.

On the other hand, *quantitative data* provides numerical information — that is, information about quantities, or amounts. For example, quantitative data on this same classroom of students can include the following:

- ✔ Each child's height in inches
- ✔ His or her weight in pounds
- ✔ The number of siblings he or she has
- ✔ The number of words he or she spelled correctly on the most recent spelling test

You can identify quantitative data by noticing that it links a number to each member of the data set. For example, Carlos is 55 inches tall, weighs 68 pounds, has three siblings, and spelled 18 words correctly.

Working with qualitative data

Qualitative data usually divides a sample into discrete chunks. As my sample — which is purely fictional — I use 25 children in Sister Elena's fifth grade class. For example, suppose all 25 children in Sister Elena's class answer the three yes-or-no questions in Table 19-1.

Table 19-1	Sister Elena's Fifth-Grade Survey	
Question	*Yes*	*No*
Are you an only child?	5	20
Do you own any pets?	14	11
Do you take the bus to school?	16	9

The students also answer the question "What is your favorite color?" with the results shown in Table 19-2.

Table 19-2	Favorite Colors in Sister Elena's Class		
Color	*Number of Students*	*Color*	*Number of Students*
Blue	8	Orange	1
Red	6	Yellow	1
Green	5	Gold	1
Purple	3		

Even though the information that each child provided is non-numerical, you can handle it numerically by counting how many students made each response and working with these numbers.

Given this information, you can now make informed statements about the students in this class just by reading the charts. For instance,

- ✔ Exactly 20 children have at least one brother or sister.

- ✔ Nine children don't take the bus to school.

- ✔ Only one child's favorite color is yellow.

Playing the percentages

You can make more-sophisticated statistical statements about qualitative data by finding out the percentage of the sample that has a specific attribute. Here's how you do so:

1. **Write a statement that includes the number of members who share that attribute and the total number in the sample.**

   Suppose you want to know what percentage of students in Sister Elena's class are only children. The chart tells you that 5 students have no siblings, and you know that 25 kids are in the class. So you can begin to answer this question as follows:

   Five out of 25 children are only children.

2. **Rewrite this statement, turning the numbers into a fraction:**

   $$\frac{\text{Number who share attribute}}{\text{Number in sample}}$$

   In the example, $\frac{5}{25}$ of the children are only children.

3. **Turn the fraction into a percent, using the method I show you in Chapter 12.**

   You find that $5 \div 25 = 0.2$, so 20% of the children are only children.

Similarly, suppose you want to find out what percentage of children take the bus to school. This time, the chart tells you that 16 children take the bus, so you can write this statement:

Sixteen out of 25 children take the bus to school.

Now rewrite the statement as follows:

$\frac{16}{25}$ of the children take the bus to school.

Finally, turn this fraction into a percent. When you find 16 ÷ 25, you get 0.64, or 64%:

64% of the children take the bus to school.

Getting into the mode

The *mode* tells you the most popular answer to a statistical question. For example, in the poll of Sister Elena's class (see Tables 19-1 and 19-2), the mode groups are children who

- ✔ Have at least one brother or sister (20 students)
- ✔ Own at least one pet (14 students)
- ✔ Take the bus to school (16 students)
- ✔ Chose blue as their favorite color (8 students)

When a question divides a data set into two parts (as with all yes-or-no questions), the mode group represents more than half of the data set. But when a question divides a data set into more than two parts, the mode doesn't necessarily represent more than half of the data set.

For example, 14 children own at least one pet, and the other 11 children don't own one. So the mode group — children who own a pet — is more than half the class. But 8 of the 25 children chose blue as their favorite color. So even though this is the mode group, fewer than half the class chose this color.

With a small sample, you may have more than one mode — for example, perhaps the number of students who like red is equal to the number who like blue. However, getting multiple modes isn't usually an issue with a larger sample because it becomes less likely that exactly the same number of people will have the same preference.

Working with quantitative data

Quantitative data assigns a numerical value to each member of the sample. As my sample — again, fictional — I use five members of Sister Elena's basketball team. Suppose that the information in Table 19-3 has been gathered about each team member's height and most recent spelling test.

Table 19-3	Height and Spelling Test Scores	
Student	**Height in Inches**	**Number of Words Spelled Correctly**
Carlos	55	18
Dwight	60	20
Patrick	59	14
Tyler	58	17
William	63	18

In this section, I show you how to use this information to find the mean and median for both sets of data. Both terms refer to ways to calculate the average value in a quantitative data set. An *average* gives you a general idea of where most individuals in a data set fall so you know what kinds of results are standard. For example, the average height of Sister Elena's fifth grade class is probably less than the average height of the Los Angeles Lakers. As I show you in the sections that follow, an average can be misleading in some cases, so knowing when to use the mean versus the median is important.

Finding the mean

The mean is the most commonly used average. In fact, when most people use the word average, they're referring to the mean. Here's how you find the *mean* of a set of data:

1. **Add up all the numbers in that set.**

   For example, to find the average height of the five team members, first add up all their heights:

   $$55 + 60 + 59 + 58 + 63 = 295$$

2. **Divide this result by the total number of members in that set.**

   Divide 295 by 5 (that is, by the total number of boys on the team):

   $$295 \div 5 = 59$$

   So the mean height of the boys on Sister Elena's team is 59 inches.

Similarly, to find the mean number of words that the boys spelled correctly, first add up the number of words they spelled correctly:

$$18 + 20 + 14 + 17 + 18 = 87$$

Now divide this result by 5:

$87 \div 5 = 17.4$

As you can see, when you divide, you end up with a decimal in your answer. If you round to the nearest whole word, the mean number of words that the five boys spelled correctly is about 17 words. (For more information about rounding, see Chapter 2.)

The mean can be misleading when you have a strong *skew* in data — that is, when the data has many low values and a few very high ones, or vice versa.

For example, suppose that the president of a company tells you, "The average salary in my company is $200,000 a year!" But on your first day at work, you find out that the president's salary is $19,010,000 and each of his 99 employees earns $10,000. To find the mean, add up the total salaries:

$19,010,000 + (99 \cdot \$10,000) = \$20,000,000$

Now divide this number by the total number of people who work there:

$20,000,000 \div 100 = \$200,000$

So the president didn't lie. However, the skew in salaries resulted in a misleading mean.

Finding the median

When data values are skewed (when a few very high or very low numbers differ significantly from the rest of the data), the median can give you a more accurate picture of what's standard. Here's how to find the *median* of a set of data:

1. **Arrange the set from lowest to highest.**

   To find the median height of the boys in Table 19-3, arrange their five heights in order from lowest to highest.

   55   58   <u>59</u>   60   63

2. **Choose the middle number.**

   The middle value, 59 inches, is the median average height.

To find the median number of words that the boys spelled correctly (see Table 19-3), arrange their scores in order from lowest to highest:

14   17   <u>18</u>   18   20

This time, the middle value is 18, so 18 is the median score.

If you have an even number of values in the data set, put the numbers in order and find the mean of the *two middle numbers* in the list (see the preceding section for details on the mean). For instance, consider the following:

> 2   3   <u>5</u>   <u>7</u>   9   11

The two center numbers are 5 and 7. Add them together to get 12 and then divide by 2 to get their mean. The median in this list is 6.

Now recall the company president who makes $19,010,000 a year and his 99 employees who each earn $10,000. Here's how this data looks:

> 10,000   10,000   10,000   ...   10,000   19,010,000

As you can see, if you were to write out all 100 salaries, the center numbers would obviously both be 10,000. The median salary is $10,000, and this result is much more reflective of what you'd probably earn if you were to work at this company.

Looking at Likelihoods: Basic Probability

Probability is the mathematics of deciding how likely an event is to occur. For example,

- What's the likelihood that the lottery ticket I bought will win?
- What's the likelihood that my new car will need repairs before the warranty runs out?
- What's the likelihood that more than 100 inches of snow will fall in Manchester, New Hampshire, this winter?

Probability has a wide variety of applications in insurance, weather prediction, biological sciences, and even physics.

The study of probability started hundreds of years ago when a group of French noblemen began to suspect that math could help them turn a profit, or at least not lose so heavily, in the gambling halls that they frequented.

You can read all about the details of probability in *Probability For Dummies* (Wiley). In this section, I give you a little taste of this fascinating subject.

Figuring the probability

The *probability* that an event will occur is a fraction whose numerator (top number) and denominator (bottom number) are as follows (for more on fractions, flip to Chapter 9):

$$\frac{\text{Number of favorable outcomes}}{\text{Total number of possible outcomes}}$$

In this case, a *favorable outcome* is simply an outcome in which the event you're examining *does* happen. In contrast, a *possible outcome* is an outcome that *can* happen.

For example, suppose you want to know the probability that a tossed coin will land heads up. Notice that there are two possible outcomes (heads or tails), but only one of these outcomes is favorable — the outcome in which heads comes up. To find the probability of this event, make a fraction as follows:

$$\frac{\text{Number of favorable outcomes}}{\text{Total number of outcomes}} = \frac{1}{2}$$

So the probability that the coin will land heads up is ½.

So what's the probability that when you roll a die, the number 3 will land face up? To figure this one out, notice that there are *six* possible outcomes (1, 2, 3, 4, 5, or 6), but in only *one* of these does 3 land face up. To find the probability of this outcome, make a fraction as follows:

$$\frac{\text{Number of favorable outcomes}}{\text{Total number of outcomes}} = \frac{1}{6}$$

So the probability that the number 3 will land face up is ⅙.

And what's the probability that if you pick a card at random from a deck, it'll be an ace? To figure this out, notice that there are *52* possible outcomes (one for each card in the deck), but in only *four* of these do you pick an ace. So

$$\frac{\text{Number of favorable outcomes}}{\text{Total number of outcomes}} = \frac{4}{52}$$

So the probability that you'll pick an ace is 4/52, which reduces to 1/13 (see Chapter 9 for more on reducing fractions).

Probability is always a fraction or decimal from 0 to 1. When the probability of an outcome is 0, the outcome is *impossible*. When the probability of an outcome is 1, the outcome is *certain*.

Oh, the possibilities! Counting outcomes with multiple coins and dice

Although the basic probability formula isn't difficult, sometimes finding the numbers to plug into it can be tricky. One source of confusion is in counting the number of outcomes, both favorable and possible. In this section, I focus on tossing coins and rolling dice.

Tossing coins

When you flip a coin, you can generally get two possible outcomes: heads or tails. When you flip two coins at the same time — say, a penny and a nickel — you can get four possible outcomes:

Outcome	Penny	Nickel
#1	Heads	Heads
#2	Heads	Tails
#3	Tails	Heads
#4	Tails	Tails

When you flip three coins at the same time — say, a penny, a nickel, and a dime — eight outcomes are possible:

Outcome	Penny	Nickel	Dime
#1	Heads	Heads	Heads
#2	Heads	Heads	Tails
#3	Heads	Tails	Heads
#4	Heads	Tails	Tails
#5	Tails	Heads	Heads
#6	Tails	Heads	Tails
#7	Tails	Tails	Heads
#8	Tails	Tails	Tails

Notice the pattern: Every time you add an additional coin, the number of possible outcomes doubles. So if you flip six coins, here's how many possible outcomes you have:

$$2 \cdot 2 \cdot 2 \cdot 2 \cdot 2 \cdot 2 = 64$$

The number of possible outcomes equals the number of outcomes per coin (2) raised to the number of coins (6): Mathematically, you have $2^6 = 64$.

Here's a handy formula for calculating the number of outcomes when you're flipping, shaking, or rolling multiple coins, dice, or other objects at the same time:

Number of outcomes per object$^{\text{Number of objects}}$

Suppose you want to find the probability that six tossed coins will all fall heads up. To do this, you want to build a fraction, and you already know that the denominator — the number of possible outcomes — is 64. Only one outcome is favorable, so the numerator is 1:

$$\frac{\text{Number of favorable outcomes}}{\text{Total number of outcomes}} = \frac{1}{64}$$

So the probability that six tossed coins will all fall heads up is ⅟₆₄.

Here's a more subtle question: What's the probability that exactly five out of six tossed coins will all fall heads up? Again, you're building a fraction, and you already know that the denominator is 64. To find the numerator (favorable outcomes), think about it this way: If the first coin falls tails up, then all the rest must fall heads up. If the second coin falls tails up, then again all the rest must fall heads up. This is true of all six coins, so you have *six* favorable outcomes:

$$\frac{\text{Number of favorable outcomes}}{\text{Total number of outcomes}} = \frac{6}{64}$$

Therefore, the probability that exactly five out of six coins will fall heads up is ⁶⁄₆₄, which reduces to ³⁄₃₂ (see Chapter 9 for more on reducing fractions).

Rolling the dice

When you roll a single die, you can get six possible outcomes: 1, 2, 3, 4, 5, or 6. However, when you roll two dice, this number jumps to 36. See Figure 19-1.

Every time you add an additional die, the number of possible outcomes is multiplied by 6. So if you roll four dice, here's the number of possible outcomes:

$$6^4 = 6 \cdot 6 \cdot 6 \cdot 6 = 1,296$$

Suppose you want to calculate the possibility of rolling four 6s. The probability is a fraction, and you already know that the denominator of this fraction is 1,296. In this case, only one outcome — all four dice coming up 6 — is favorable, so here's how you build your fraction:

$$\frac{\text{Number of favorable outcomes}}{\text{Total number of outcomes}} = \frac{1}{1,296}$$

So the probability that you'll roll four 6s is $\frac{1}{1,296}$ — a very small probability, indeed.

Here's a more interesting question: What's the probability that all four dice will come up 4, 5, or 6? Again, you're building a fraction whose denominator is 1,296. To find the numerator, think about it this way: For the first die, there are three favorable outcomes (4, 5, or 6). For the first two dice, there are $3 \cdot 3 = 9$ favorable outcomes as shown here:

4-4	4-5	4-6
5-4	5-5	5-6
6-4	6-5	6-6

For three dice, there are $3 \cdot 3 \cdot 3 = 27$ favorable outcomes. So for all four dice, there are $3 \cdot 3 \cdot 3 \cdot 3 = 81$ favorable outcomes. So

$$\frac{\text{Number of favorable outcomes}}{\text{Total number of outcomes}} = \frac{3^4}{6^4} = \frac{81}{1,296}$$

Thus, the probability that all four dice will come up 4, 5, or 6 is $\frac{81}{1,296}$. This fraction reduces to $\frac{1}{16}$ (see Chapter 9 for more on reducing fractions).

Chapter 20

Setting Things Up
with Basic Set Theory

In This Chapter
▶ Defining a set and its elements

▶ Understanding subsets and the empty set

▶ Knowing the basic operations on sets, including union and intersection

A *set* is just a collection of things. But in their simplicity, sets are profound. At the deepest level, set theory is the foundation for everything in math.

Set theory provides a way to talk about collections of numbers such as even numbers, prime numbers, or counting numbers with ease and clarity. It also gives rules for performing calculations on sets that become useful in higher math. For these reasons, set theory becomes more important the higher you go up the math food chain — especially when you begin writing mathematical proofs. Studying sets can also be a nice break from the usual math stuff you work with.

In this chapter, I show you the basics of set theory. First, I show you how to define sets and their elements and how you can tell when two sets are equal. I also show you the simple idea of a set's cardinality. Next, I discuss subsets and the all-important empty set (∅). After that, I discuss four operations on sets: union, intersection, relative complement, and complement.

Understanding Sets

A *set* is a collection of things, in any order. These things can be buildings, earmuffs, lightning bugs, numbers, qualities of historical figures, names you call your little brother, whatever.

You can define a set in a few main ways:

- **Placing a list of the elements of the set in braces { }:** You can simply list everything that belongs in the set. When the set is too large, simply use an ellipsis (...) to indicate elements of the set not mentioned. For example, to list the set of numbers from 1 to 100, you can write {1, 2, 3, ..., 100}. To list the set of *all* the counting numbers, you can write {1, 2, 3, ...}.

- **Using a verbal description:** If you use a verbal description of what the set includes, make sure the description is clear and unambiguous so you know exactly what's in the set and what isn't. For instance, the set of the four seasons is pretty clear-cut, but you may run into some debate on the set of words that describe my cooking skills because different people have different opinions.

- **Writing a mathematical rule (set-builder notation):** In later algebra, you can write an equation that tells people how to calculate the numbers that are part of a set. Check out *Algebra II For Dummies,* by Mary Jane Sterling (Wiley), for details.

Sets are usually identified with capital letters to keep them distinct from variables in algebra, which are usually small letters. (Chapter 21 talks about using variables.)

The best way to understand sets is to begin working with them. For example, here I define three sets:

A = {Empire State Building, Eiffel Tower, Roman Colosseum}

B = {Albert Einstein's intelligence, Marilyn Monroe's talent, Joe DiMaggio's athletic ability, Senator Joseph McCarthy's ruthlessness}

C = the four seasons of the year

Set A contains three tangible objects: famous works of architecture. Set B contains four intangible objects: attributes of famous people. And set C also contains intangible objects: the four seasons. Set theory allows you to work with either tangible or intangible objects, provided that you define your set properly. In the following sections, I show you the basics of set theory.

Elementary, my dear: Considering what's inside sets

The things contained in a set are called *elements* (also known as *members*). Consider the first two sets I define in the section intro:

A = {Empire State Building, Eiffel Tower, Roman Colosseum}

B = {Albert Einstein's intelligence, Marilyn Monroe's talent, Joe
DiMaggio's athletic ability, Senator Joseph McCarthy's ruthlessness}

The Eiffel Tower is an element of A, and Marilyn Monroe's talent is an element of B. You can write these statements using the symbol $\in$, which means *is an element of:*

Eiffel Tower $\in$ A

Marilyn Monroe's talent $\in$ B

However, the Eiffel Tower is not an element of B. You can write this statement using the symbol $\notin$, which means *is not an element of:*

Eiffel Tower $\notin$ B

These two symbols become more common as you move higher in your study of math. The following sections discuss what's inside those braces and how some sets relate to each other.

Cardinality of sets

The *cardinality* of a set is just a fancy word for the number of elements in that set.

When A is {Empire State Building, Eiffel Tower, Roman Colosseum}, it has three elements, so the cardinality of A is three. Set B, which is {Albert Einstein's intelligence, Marilyn Monroe's talent, Joe DiMaggio's athletic ability, Senator Joseph McCarthy's ruthlessness}, has four elements, so the cardinality of B is four.

Equal sets

If two sets list or describe the exact same elements, the sets are *equal* (you can also say they're *identical* or *equivalent*). The order of elements in the sets doesn't matter. Similarly, an element may appear twice in one set, but only the distinct elements need to match.

Suppose I define some sets as follows:

C = the four seasons of the year

D = {spring, summer, fall, winter}

E = {fall, spring, summer, winter}

F = {summer, summer, summer, spring, fall, winter, winter, summer}

Set C gives a clear rule describing a set. Set D explicitly lists the four elements in C. Set E lists the four seasons in a different order. And set F lists the four seasons with some repetition. Thus, all four sets are *equal*. As with numbers, you can use the equal sign to show that sets are equal:

C = D = E = F

Subsets

When all the elements of one set are completely contained in a second set, the first set is a *subset* of the second. For example, consider these sets:

C = {spring, summer, fall, winter}

G = {spring, summer, fall}

As you can see, every element of G is also an element of C, so G is a subset of C. The symbol for subset is ⊂, so you can write the following:

G ⊂ C

Every set is a subset of itself. This idea may seem odd until you realize that all the elements of any set are obviously contained in that set.

Empty sets

The *empty set* — also called the *null set* — is a set that has no elements:

H = {}

As you can see, I define H by listing its elements, but I haven't listed any, so H is empty. The symbol ⊘ is used to represent the empty set. So H = ⊘.

You can also define an empty set using a rule. For example,

I = types of roosters that lay eggs

Clearly, roosters are male and therefore can't lay eggs, so this set is empty.

You can think of ⊘ as *nothing*. And because *nothing* is always *nothing,* there's only one empty set. All empty sets are equal to each other, so in this case, H = I.

Furthermore, ⊘ is a subset of every other set (the preceding section discusses subsets), so the following statements are true:

⊘ ⊂ A

⊘ ⊂ B

⊘ ⊂ C

This concept makes sense when you think about it. Remember that ∅ has *no* elements, so technically speaking, *every* element in ∅ is in every other set.

Sets of numbers

One important use of sets is to define sets of numbers. As with all other sets, you can do so either by listing the elements or by verbally describing a rule that clearly tells you what's included in the set and what isn't. For example, consider the following sets:

J = {1, 2, 3, 4, 5}

K = {2, 4, 6, 8, 10, ...}

L = the set of counting numbers

My definitions of J and K list their elements explicitly. Because K is infinitely large, I need to use an ellipsis (...) to show that this set goes on forever. The definition of L is a description of the set in words.

I discuss some especially significant sets of numbers in Chapter 25.

Operations on Sets

In arithmetic, the Big Four operations (adding, subtracting, multiplying, and dividing) allow you to combine numbers in various ways (see Chapters 3 and 4 for more information). Set theory also has four important operations: union, intersection, relative complement, and complement. You'll see more of these operations as you move on in your study of math.

Here are definitions for three sets of numbers:

P = {1, 7}

Q = {4, 5, 6}

R = {2, 4, 6, 8, 10}

In this section, I use these three sets and a few others to discuss the four set operations and show you how they work. (***Note:*** Within equations, I re-list the elements, replacing the names of the sets with their equivalent in braces. Therefore, you shouldn't have to flip back and forth to look up what each set contains.)

Union: Combined elements

The union of two sets is the set of their *combined* elements. For example, the union of {1, 2} and {3, 4} is {1, 2, 3, 4}. The symbol for this operation is ∪, so

{1, 2} ∪ {3, 4} = {1, 2, 3, 4}

Similarly, here's how to find the union of P and Q:

P ∪ Q = {1, 7} ∪ {4, 5, 6} = {1, 4, 5, 6, 7}

When two sets have one or more elements in common, these elements appear only once in their union set. For example, consider the union of Q and R. In this case, the elements 4 and 6 are in both sets, but each of these numbers appears once in their union:

Q ∪ R = {4, 5, 6} ∪ {2, 4, 6, 8, 10} = {2, 4, 5, 6, 8, 10}

The union of any set with itself is itself:

P ∪ P = P

Similarly, the union of any set with ∅ (see the earlier "Empty sets" section) is itself:

P ∪ ∅ = P

Intersection: Elements in common

The *intersection* of two sets is the set of their common elements (the elements that appear in both sets). For example, the intersection of {1, 2, 3} and {2, 3, 4} is {2, 3}. The symbol for this operation is ∩. You can write the following:

{1, 2, 3} ∩ {2, 3, 4} = {2, 3}

Similarly, here's how to write the intersection of Q and R:

Q ∩ R = {4, 5, 6} ∩ {2, 4, 6, 8, 10} = {4, 6}

When two sets have no elements in common, their intersection is the empty set (∅):

P ∩ Q = {1, 7} ∩ {4, 5, 6} = ∅

The intersection of any set with itself is itself:

$P \cap P = P$

But the intersection of any set with $\oslash$ is $\oslash$:

$P \cap \oslash = \oslash$

Relative complement: Subtraction (sorta)

The *relative complement* of two sets is an operation similar to subtraction. The symbol for this operation is the minus sign (–). Starting with the first set, you remove every element that appears in the second set to arrive at their relative complement. For example,

$\{1, 2, 3, 4, 5\} - \{1, 2, 5\} = \{3, 4\}$

Similarly, here's how to find the relative complement of R and Q. Both sets share a 4 and a 6, so you have to remove those elements from R:

$R - Q = \{2, 4, 6, 8, 10\} - \{4, 5, 6\} = \{2, 8, 10\}$

Note that the reversal of this operation gives you a different result. This time, you remove the shared 4 and 6 from Q:

$Q - R = \{4, 5, 6\} - \{2, 4, 6, 8, 10\} = \{5\}$

Like subtraction in arithmetic, the relative complement is *not* a commutative operation. In other words, order is important. (See Chapter 4 for more on commutative and noncommutative operations.)

Complement: Feeling left out

The *complement* of a set is everything that isn't in that set. Because *everything* is a difficult concept to work with, you first have to define what you mean by *everything* as the *universal set* (U). For example, suppose you define the universal set like this:

$U = \{0, 1, 2, 3, 4, 5, 6, 7, 8, 9\}$

Now, here are a couple of sets to work with:

$M = \{1, 3, 5, 7, 9\}$

$N = \{6\}$

The complement of each set is the set of every element in U that isn't in the original set:

$U - M = \{0, 1, 2, 3, 4, 5, 6, 7, 8, 9\} - \{1, 3, 5, 7, 9\} = \{0, 2, 4, 6, 8\}$

$U - N = \{0, 1, 2, 3, 4, 5, 6, 7, 8, 9\} - \{6\} = \{0, 1, 2, 3, 4, 5, 7, 8, 9\}$

The complement is closely related to the relative complement (see the preceding section). Both operations are similar to subtraction. The main difference is that the complement is *always* subtraction of a set from U, but the relative complement is subtraction of a set from any other set.

The symbol for the complement is ', so you can write the following:

$M' = \{0, 2, 4, 6, 8\}$

$N' = \{0, 1, 2, 3, 4, 5, 7, 8, 9\}$

Part V
The X-Files: Introduction to Algebra

The 5th Wave   By Rich Tennant

©RICHTENNANT

"David's using algebra to calculate the tip. Barbara— would you mind being a fractional exponent?"

In this part . . .

1 introduce you to algebra, an amazing tool for solving a wide variety of math problems with speed and accuracy. I show you how to declare a variable (such as x) to stand for an unknown number. Then, I give you a bunch of important tools for working with algebraic expressions. Next, you put these tools to work solving algebraic equations. Finally, everything comes together as you solve algebra word problems.

Chapter 21

Enter Mr. X: Algebra and Algebraic Expressions

*Y*ou never forget your first love, your first car, or your first *x*. Unfortunately for some folks, remembering their first *x* in algebra is similar to remembering their first love who stood them up at the prom or their first car that broke down someplace in Mexico.

The most well-known fact about algebra is that it uses letters — like *x* — to represent numbers. So, if you have a traumatic *x*-related tale, all I can say is that the future will be brighter than the past.

What good is algebra? That's a common question, and it deserves a decent answer. Algebra is used for solving problems that are just too difficult for ordinary arithmetic. And because number crunching is so much a part of the modern world, algebra is everywhere (even if you don't see it): architecture, engineering, medicine, statistics, computers, business, chemistry, physics, biology, and of course higher math. Anywhere that numbers are useful, algebra is there. That's why virtually every college and university insists that you leave (or enter) with at least a passing familiarity with algebra.

In this chapter, I introduce (or re-introduce) you to that elusive little fellow, Mr. *X*, in a way that's bound to make him seem a little friendlier. Then I show you how *algebraic expressions* are similar to and different from the arithmetic expressions that you're used to working with. (For a refresher on arithmetic expressions, see Chapter 5.)

X Marks the Spot

In math, x stands for a number — any number. Any letter that you use to stand for a number is a *variable,* which means that its value can *vary* — that is, its value is uncertain. In contrast, a number in algebra is often called a *constant,* because its value is fixed.

Sometimes, you have enough information to find out the identity of x. For example, consider the following:

$$2 + 2 = x$$

Obviously, in this equation, x stands for the number 4. But other times, the number that x stands for stays shrouded in mystery. For example:

$$x > 5$$

In this inequality, x stands for some number greater than 5 — maybe 6, maybe 7½, maybe 542.002.

Expressing Yourself with Algebraic Expressions

In Chapter 5, I introduce you to *arithmetic expressions:* strings of numbers and operators that can be evaluated or placed on one side of an equation. For example:

$$2 + 3$$
$$7 \cdot 1.5 - 2$$
$$2^4 - |-4| - \sqrt{100}$$

In this chapter, I introduce you to another type of mathematical expression: the algebraic expression. An *algebraic expression* is any string of mathematical symbols that can be placed on one side of an equation and that includes at least one variable.

Here are a few examples of algebraic expressions:

$$5x$$

$$-5x + 2$$

$$x^2y - y^2x + \frac{z}{3} - xyz + 1$$

As you can see, the difference between arithmetic and algebraic expressions is simply that an algebraic expression includes at least one variable.

In this section, I show you how to work with algebraic expressions. First, I demonstrate how to evaluate an algebraic expression by substituting the values of its variables. Then, I show you how to separate an algebraic expression into one or more terms and how to identify the coefficient and the variable part of each term.

Evaluating algebraic expressions

To evaluate an algebraic expression, you need to know the numerical value of every variable. For each variable in the expression, *substitute* the number that it stands for, then evaluate the expression.

In Chapter 5, I show you how to evaluate an arithmetic expression. Briefly, this means finding the value of that expression as a single number (flip to Chapter 5 for more on evaluating).

Knowing how to evaluate arithmetic expressions comes in handy for evaluating algebraic expressions. For example, suppose you want to evaluate the following expression:

$$4x - 7$$

Note that this expression contains the variable x, which is unknown, so the value of the whole expression is also unknown.

An algebraic expression can have any number of variables, but you usually don't work with expressions that have more than three. You can use any letter as a variable, but x, y, and z tend to get a lot of mileage.

Suppose in the previous case that $x = 2$. To evaluate the expression, *substitute* 2 for x everywhere it appears in the expression:

$4(2) - 7$

After you make the substitution, you're left with an arithmetic expression, so you can finish your calculations to evaluate the expression:

$= 8 - 7 = 1$

So given $x = 2$, the algebraic expression $4x - 7 = 1$.

Now suppose you want to evaluate the following expression, where $x = 4$:

$2x^2 - 5x - 15$

This time, the first step is to substitute 4 for x everywhere that this variable appears in the expression:

$2(4^2) - 5(4) - 15$

Now evaluate according to the order of operations explained in Chapter 5. You do powers first, so begin by evaluating the exponent 4^2, which equals $4 \cdot 4$:

$= 2(16) - 5(4) - 15$

Now proceed on to the multiplication, moving from left to right:

$= 32 - 5(4) - 15$

$= 32 - 20 - 15$

Then evaluate the subtraction, again from left to right:

$= 12 - 15$

$= -3$

So given $x = 4$, the algebraic expression $2x^2 - 5x - 15 = -3$.

You aren't limited to expressions of only one variable when using substitution. As long as you know the value of every variable in the expression, you can evaluate algebraic expressions with any number of variables. For example, suppose you want to evaluate this expression:

$3x^2 + 2xy - xyz$

To evaluate it, you need values of all three variables:

$x = 3$

$y = -2$

$z = 5$

The first step is to substitute the equivalent value for each of the three variables wherever you find them:

$3(3^2) + 2(3)(-2) - 3(-2)(5)$

Now use the rules for order of operations from Chapter 5. Begin by evaluating the exponent 3^2:

$= 3(9) + 2(3)(-2) - 3(-2)(5)$

Next, evaluate the multiplication from left to right (if you need to know more about the rules for multiplying negative numbers, check out Chapter 4):

$-27 + 2(3)(-2) - 3(-2)(5)$

$= 27 + 6(-2) - 3(-2)(5)$

$= 27 + -12 - 3(-2)(5)$

$= 27 + -12 - (-6)(5)$

$= 27 + -12 - (-30)$

Now all that's left is addition and subtraction. Evaluate from left to right, remembering the rules for adding and subtracting negative numbers in Chapter 4:

$= 15 - (-30) = 15 + 30 = 45$

So given the three values for x, y, and z, the algebraic expression $3x^2 + 2xy - xyz = 45$.

For practice, copy down the expression and the three values, close the book, and see whether you can substitute and evaluate on your own.

Coming to algebraic terms

A *term* in an algebraic expression is any chunk of symbols set off from the rest of the expression by either addition or subtraction. As algebraic expressions get more complex, they begin to string themselves out in more and more terms. Here are some examples:

Expression	Number of Terms	Terms
$5x$	One	$5x$
$-5x + 2$	Two	$-5x$ and 2
$x^2y + \frac{z}{3} - xyz + 8$	Four	$x^2y, \frac{z}{3}, -xyz,$ and 8

No matter how complicated an algebraic expression gets, you can always separate it out into one or more terms.

When separating an algebraic expression into terms, group the plus or minus sign with the term that it immediately precedes.

When a term has a variable, it's called an *algebraic term*. When it doesn't have a variable, it's called a *constant*. For example, look at the following expression:

$$x^2y + \frac{z}{3} - xyz + 8$$

The first three terms are algebraic terms, and the last term is a constant. As you can see, in algebra, *constant* is just a fancy word for *number*.

Terms are really useful because you can follow rules to move them, combine them, and perform the Big Four operations on them. All these skills are important for solving equations, which I explain in the next chapter. But for now, this section explains a bit about terms and some of their traits.

Making the commute: Rearranging your terms

After you understand how to separate an algebraic expression into terms, you can go one step further by rearranging the terms in any order you like. Each term moves as a unit, kind of like a group of people carpooling to work together — everyone in the car stays together for the whole ride.

For example, suppose you begin with the expression $-5x + 2$. You can rearrange the two terms of this expression without changing its value. Notice that each term's sign stays with that term, though dropping the plus sign at the beginning of an expression is customary:

$$= 2 - 5x$$

Rearranging terms in this way doesn't affect the value of the expression because addition is commutative — that is, you can rearrange things that you're adding and without changing the answer. (See Chapter 4 for more on the commutative property of addition.)

For example, suppose $x = 3$. Then the original expression and its rearrangement evaluate as follows (using the rules that I outline earlier in "Evaluating Algebraic Expressions"):

$$-5x + 2 \qquad\qquad 2 - 5x$$
$$= -5(3) + 2 \qquad = 2 - 5(3)$$
$$= -15 + 2 \qquad\quad = 2 - 15$$
$$= -13 \qquad\qquad\; = -13$$

Rearranging expressions in this way becomes handy later in this chapter, when you simplify algebraic expressions. As another example, suppose you have this expression:

$$4x - y + 6$$

You can rearrange it in a variety of ways:

$$= 6 + 4x - y$$
$$= -y + 4x + 6$$

Because the term $4x$ has no sign, it's positive, so you can write in a plus sign as needed when rearranging terms.

As long as each term's sign stays with that term, rearranging the terms in an expression has no effect on its value.

For example, suppose that $x = 2$ and $y = 3$. Here's how to evaluate the original expression and the two re-arrangements:

$$4x - y + 6 \qquad\quad 6 + 4x - y \qquad\quad -y + 4x + 6$$
$$= 4(2) - 3 + 6 \quad = 6 + 4(2) - 3 \quad = -3 + 4(2) + 6$$
$$= 8 - 3 + 6 \qquad = 6 + 8 - 3 \qquad = -3 + 8 + 6$$
$$= 5 + 6 \qquad\qquad = 14 - 3 \qquad\quad = 5 + 6$$
$$= 11 \qquad\qquad\;\; = 11 \qquad\qquad\;\; = 11$$

Identifying the coefficient and variable

Every term in an algebraic expression has a coefficient. The *coefficient* is the signed numerical part of a term in an algebraic expression — that is, the number and the sign (+ or -) that goes with that term. For example, suppose you're working with the following algebraic expression:

$$-4x^3 + x^2 - x - 7$$

The table below shows the four terms of this expression, with each term's coefficient:

Term	Coefficient	Variable
$-4x^3$	-4	x^3
x^2	1	x^2
$-x$	-1	x
-7	-7	none

Notice that the *sign* associated with the term is part of the coefficient. So the coefficient of $-4x^3$ is -4.

When a term appears to have no coefficient, the coefficient is actually 1. So the coefficient of x^2 is 1, and the coefficient of $-x$ is -1. And when a term is a constant (just a number), that number with its associated sign is the coefficient. So the coefficient of the term -7 is simply -7.

By the way, when the coefficient of any algebraic term is 0, the expression equals 0 no matter what the variable part looks like:

$$0x = 0 \qquad 0xyz = 0 \qquad 0x^3y^4z^{10} = 0$$

In contrast, the *variable part* of an expression is everything except the coefficient. The table above shows the four terms of the same expression, with each term's variable part.

Identifying similar terms

Similar terms (or *like terms*) are any two algebraic terms that have the same variable part — that is, both the letters and their exponents have to be exact matches. Here are some examples:

Variable Part	*Examples of Similar Terms*		
x	$4x$	$12x$	$99.9x$
x^2	$6x^2$	$-20x^2$	$\frac{8}{3}x^2$
y	y	$1,000y$	πy
xy	$-7xy$	$800xy$	$\frac{22}{7}xy$
x^3y^3	$3x^3y^3$	$-111x^3y^3$	$3.14x^3y^3$

As you can see, in each example, the variable part in all three similar terms is the same. Only the coefficient changes, and it can be any real number: positive or negative, whole number, fraction, decimal, or even an irrational number such as π. (For more on real numbers, see Chapter 25.)

Considering algebraic terms and the Big Four

In this section, I get you up to speed on how to apply the Big Four to algebraic expressions. For now, just think of working with algebraic expressions as a set of tools that you're collecting, for use when you get on the job. You find how useful these tools are in Chapter 22, when you begin solving algebraic equations.

Adding terms

Add similar terms by adding their coefficients and keeping the same variable part.

For example, suppose you have the expression $2x + 3x$. Remember that $2x$ is just shorthand for $x + x$, and $3x$ means simply $x + x + x$. So when you add them up, you get the following:

$$= x + x + x + x + x = 5x$$

As you can see, when the variable parts of two terms are the same, you add these terms by adding their coefficients: $2x + 3x = (2 + 3)x$. The idea here is roughly similar to the idea that 2 apples + 3 apples = 5 apples.

You *cannot* add non-similar terms. Here are some cases in which the variables or their exponents are different:

$2x + 3y$

$2yz + 3y$

$2x^2 + 3x$

In these cases, you can't perform the addition. You're faced with a situation that's similar to 2 apples + 3 oranges. Because the units (apples and oranges) are different, you can't solve the problem. (See Chapter 4 for more on how to work with units.)

Subtracting terms

Subtraction works much the same as addition. Subtract similar terms by finding the difference between their coefficients and keeping the same variable part.

For example, suppose you have $3x - x$. Recall that $3x$ is simply shorthand for $x + x + x$. So doing this subtraction gives you the following:

$$x + x + x - x = 2x$$

No big surprises here. You simply find $(3 - 1)x$. This time, the idea roughly parallels the idea that $\$3 - \$1 = \$2$.

Here's another example:

$$2x - 5x$$

Again, no problem, as long as you know how to work with negative numbers (see Chapter 4 if you need details). Just find the difference between the coefficients:

$$= (2 - 5)x = -3x$$

In this case, recall that $\$2 - \$5 = -\$3$ (that is, a debt of 3 dollars).

You *cannot* subtract non-similar terms. For example, you can't subtract either of the following:

$$7x - 4y$$

$$7x^2y - 4xy^2$$

As with addition, you can't do subtraction with different variables. Think of this as trying to figure out $\$7 - 4$ pesos. Because the units in this case (dollars versus pesos) are different, you're stuck. (See Chapter 4 for more on working with units.)

Multiplying terms

Unlike adding and subtracting, you can multiply non-similar terms. Multiply *any* two terms by multiplying their coefficients and combining — that is, by collecting or gathering up — all the variables in each term into a single term, as I show you below.

For example, suppose you want to multiply $5x(3y)$. To get the coefficient, multiply $5 \cdot 3$. To get the algebraic part, combine the variables x and y:

$$= 5(3)xy = 15xy$$

Now suppose you want to multiply $2x(7x)$. Again, multiply the coefficients and collect the variables into a single term:

$$= 7(2)xx = 14xx$$

Remember that x^2 is shorthand for xx, so you can write the answer more efficiently:

$$= 14x^2$$

Here's another example. Multiply all three coefficients together and gather up the variables:

$$2x^2(3y)(4xy)$$
$$= 2(3)(4)x^2xyy$$
$$= 24x^3y^2$$

As you can see, the exponent 3 that's associated with x is just the count of how many x's appear in the problem. The same is true of the exponent 2 associated with y.

A fast way to multiply variables with exponents is to add the exponents together. For example:

$$(x^4y^3)(x^2y^5)(x^6y) = x^{12}y^9$$

In this example, I added the exponents of the x's ($4 + 2 + 6 = 12$) to get the exponent of x in the solution. Similarly, I added the exponents of the y's ($3 + 5 + 1 = 9$ — don't forget that $y = y^1$!) to get the exponent of y in the solution.

Dividing terms

It's customary to represent division of algebraic expressions as a fraction instead of using the division sign (÷). So division of algebraic terms really looks like reducing a fraction to lowest terms (see Chapter 9 for more on reducing).

To divide one algebraic term by another, follow these steps:

1. **Make a fraction of the two terms.**

   Suppose you want to divide $3xy$ by $12x^2$. Begin by turning the problem into a fraction:

   $$\frac{3xy}{12x^2}$$

2. **Cancel out factors in coefficients that are in both the numerator and denominator.**

   In this case, you can cancel out a 3. Notice that when the coefficient in xy becomes 1, you can drop it:

   $$=\frac{xy}{4x^2}$$

3. **Cancel out any variable that's in both the numerator and denominator.**

   You can break x^2 out as xx:

   $$=\frac{xy}{4xx}$$

   Now you can clearly cancel an x in both the numerator and denominator:

   $$=\frac{y}{4x}$$

   As you can see, the resulting fraction is really a reduced form of the original.

As another example, suppose you want to divide $-6x^2yz^3$ by $-8x^2y^2z$. Begin by writing the division as a fraction:

$$\frac{-6x^2yz^3}{-8x^2y^2z}$$

First, reduce the coefficients. Notice that because both coefficients were originally negative, you can cancel out both minus signs as well:

$$=\frac{3x^2yz^3}{4x^2y^2z}$$

Now you can begin canceling variables. I do this in two steps as before:

$$= \frac{3xxyzzz}{4xxyyz}$$

At this point, just cross out any occurrence of a variable that appears in both the numerator and denominator:

$$= \frac{3zz}{4y}$$

$$= \frac{3z^2}{4y}$$

You can't cancel out variables or coefficients if either the numerator or denominator has more than one term in it.

Simplifying Algebraic Expressions

As algebraic expressions grow more complex, simplifying them can make them easier to work with. Simplifying an expression means (quite simply!) making it smaller and easier to manage. You see how important simplifying expressions becomes when you begin solving algebraic equations.

For now, think of this section as a kind of algebra toolkit. Here, I show you *how* to use these tools. In Chapter 22, I show you *when* to use them.

Combining similar terms

When two algebraic terms are similar (when their variables match), you can add or subtract them (see the earlier "Considering algebraic terms and the Big Four" section). This feature comes in handy when you're trying to simplify an expression. For example, suppose you're working with following expression:

$$4x - 3y + 2x + y - x + 2y$$

As it stands, this expression has six terms. But three terms have the variable x and the other three have the variable y. Begin by rearranging the expression so that all similar terms are grouped together:

$$= 4x + 2x - x - 3y + y + 2y$$

Now you can add and subtract similar terms. I do this in two steps, first for the x terms and then for the y terms:

$$= 5x - 3y + y + 2y$$
$$= 5x + 0y = 5x$$

Notice that the x terms simplify to $5x$ and the y terms simplify to $0y$, which is 0, so the y terms drop out of the expression altogether!

Here's a somewhat more complicated example that has variables with exponents:

$$12x - xy - 3x^2 + 8y + 10xy + 3x^2 - 7x$$

This time, you have four different types of terms. As a first step, you can rearrange these terms so that groups of similar terms are all together (I underline these four groups so you can see them clearly):

$$= \underline{12x - 7x} \ \underline{-xy + 10xy} \ \underline{-3x^2 + 3x^2} \ \underline{+8y}$$

Now combine each set of similar terms:

$$= 5x + 9xy + 0x^2 + 8y$$

This time, the x^2 terms add up to 0, so they drop out of the expression altogether:

$$= 5x + 9xy + 8y$$

Removing parentheses from an algebraic expression

Parentheses keep parts of an expression together as a single unit. In Chapter 5, I show you how to handle parentheses in an arithmetic expression. This skill is also useful with algebraic expressions. As you find when you begin solving algebraic equations in Chapter 22, getting rid of parentheses is often the first step toward solving a problem. In this section, I show how to handle the Big Four operations with ease.

Drop everything: Parentheses with a plus sign

When an expression contains parentheses that come right after a plus sign (+), you can just remove the parentheses. Here's an example:

$$2x \underline{+ (3x - y)} + 5y$$
$$= 2x \underline{+ 3x - y} + 5y$$

Now you can simplify the expression by combining similar terms:

$= 5x + 4y$

When the first term inside the parentheses is negative, when you drop the parentheses, the minus sign replaces the plus sign. For example:

$6x + (-2x + y) - 4y$

$= 6x - 2x + y - 4y$

Sign turnabout: Parentheses with a minus sign

Sometimes, an expression contains parentheses that come right after a minus sign (–). In this case, change the sign of every term inside the parentheses to the opposite sign; then remove the parentheses.

Consider this example:

$6x - (2xy - 3y) + 5xy$

A minus sign is in front of the parentheses, so you need to change the signs of both terms in the parentheses and remove the parentheses. Notice that the term $2xy$ appears to have no sign because it's the first term inside the parentheses. This expression really means the following:

$= 6x \underline{- (+2xy - 3y)} + 5xy$

You can see how to change the signs:

$= 6x \underline{- 2xy + 3y} + 5xy$

At this point, you can combine the two similar xy terms:

$= 6x + 3xy + 3y$

Distributing: Parentheses with no sign

When you see nothing between a number and a set of parentheses, it means multiplication. For example,

$2(3) = 6$ $4(4) = 16$ $10(15) = 150$

This notation becomes much more common with algebraic expressions, replacing even the dot multiplication sign (·):

$3(4x) = 12x$ $4x(2x) = 8x^2$ $3x(7y) = 21xy$

To remove parentheses without a sign, multiply the term outside the parentheses by every term inside the parentheses; then remove the parentheses. When you follow those steps, you're using the *distributive property*.

Here's an example:

$$2(3x - 5y + 4)$$

In this case, multiply 2 by each of the three terms inside the parentheses:

$$= 2(3x) + 2(-5y) + 2(4)$$

For the moment, this expression looks more complex than the original one, but now you can get rid of all three sets of parentheses by multiplying:

$$= 6x - 10y + 8$$

Multiplying by every term inside the parentheses is simply distribution of multiplication over addition — also called the *distributive property* — which I discuss in Chapter 4.

As another example, suppose you have the following expression:

$$-2x(-3x + y + 6) + 2xy - 5x^2$$

Begin by multiplying $-2x$ by the three terms inside the parentheses:

$$= -2x(-3x) - 2x(y) - 2x(6) + 2xy - 5x^2$$

The expression looks worse than when you started, but you can get rid of all the parentheses by multiplying:

$$= 6x^2 - 2xy - 12x + 2xy - 5x^2$$

Now you can combine similar terms. I do this in two steps, first rearranging and then combining:

$$= 6x^2 - 5x^2 - 2xy + 2xy - 12x$$
$$= x^2 - 12x$$

Parentheses by FOILing

Sometimes, expressions have two sets of parentheses next to each other without a sign between them. In that case, you need to multiply *every term* inside the first set by *every term* inside the second.

When you have two terms inside each set of parentheses, you can use a process called FOILing. The word *FOIL* is an acronym to help you make sure you multiply the correct terms. It stands for *First, Outside, Inside,* and *Last.* Here's how the process works:

1. **Start out by multiplying the two *First* terms in the parentheses.**

   Suppose you want to simplify the expression $(2x - 2)(3x - 6)$. The first term in the first set of parentheses is $2x$, and $3x$ is the first term in the second set of parentheses. Therefore, multiply $2x$ by $3x$:

   $$(\underline{2x} - 2)(\underline{3x} - 6) \quad 2x(3x) = 6x^2$$

2. **Then multiply the two *Outside* terms.**

   The two outside terms, $2x$ and -6, are on the ends:

   $$(\underline{2x} - 2)(3x \underline{- 6}) \quad 2x(-6) = -12x$$

3. **Next, multiply the two *Inside* terms.**

   The two terms in the middle are -2 and $3x$:

   $$(2x \underline{- 2})(\underline{3x} - 6) \quad -2(3x) = -6x$$

4. **Finally, multiply the two *Last* terms.**

   The last term in the first set of parentheses is -2, and -6 is last term in the second set:

   $$(2x \underline{- 2})(3x \underline{- 6}) \quad -2(-6) = 12$$

Add these four results together to get the simplified expression:

$$6x^2 - 12x - 6x + 12$$

In this case, you can simplify this expression still further by combining the similar terms $-12x$ and $-6x$:

$$= 6x^2 - 18x + 12$$

Notice that during this process, you multiply every term inside one set of parentheses by every term inside the other set. FOILing just helps you keep track and make sure you've multiplied everything.

FOILing is really just an application of the distributive property, which I discuss in the section before this one. In other words, $(2x - 2)(3x - 6)$ is really the same thing as $2x(3x - 6) + -2(3x - 6)$ when distributed. Then, distributing again gives you $6x^2 - 6x - 12x + 12$.

Chapter 22

Unmasking Mr. X: Algebraic Equations

In This Chapter

▶ Using variables (such as x) in equations

▶ Knowing some quick ways to solve for x in simple equations

▶ Understanding the *balance-scale method* for solving equations

▶ Rearranging terms in an algebraic equation

▶ Isolating algebraic terms on one side of an equation

▶ Removing parentheses from an equation

▶ Cross multiplying to remove fractions

---

*W*hen it comes to algebra, solving equations is the main event.

Solving an algebraic equation means finding out what number the variable (usually x) stands for. Not surprisingly, this process is called *solving for x,* and when you know how to do it, your confidence — not to mention your grades — in your algebra class will soar through the roof.

That's what this chapter is all about. First, I show you a few informal methods to solve for x when an equation isn't too difficult. Then, I show you how to solve more difficult equations by thinking of them as a balance scale.

The balance-scale method is really the heart of algebra (yes, algebra has a heart after all!). After you understand this simple idea, you're ready to solve more complicated equations using all the tools I show you in Chapter 21, such as simplifying expressions and removing parentheses. You find out how to extend these skills to algebraic equations. Finally, I show you how cross multiplying (see Chapter 9) can make solving algebraic equations with fractions a piece of cake.

By the end of this chapter, you should have a solid grasp of a bunch of ways to solve equations for the elusive and mysterious x.

Understanding Algebraic Equations

An algebraic equation is an equation that includes at least one variable — that is, a letter (such as x) that stands for a number. *Solving* an algebraic equation means finding out what number x stands for.

In this section, I show you the basics of how a variable like x works its way into an equation in the first place. Then I show you a few quick ways to *solve for x* when an equation isn't too difficult.

Using x in equations

As you discover in Chapter 5, an *equation* is a mathematical statement that contains an equal sign. For example, here's a perfectly good equation:

$$7 \cdot 9 = 63$$

At its heart, a variable (such as x) is nothing more than a placeholder for a number. You're probably used to equations that use other placeholders: One number is purposely left as a blank or replaced by an underline or a question mark, and you're supposed to fill it in. Usually, this number comes after the equal sign. For example:

$$8 + 2 =$$
$$12 - 3 = \underline{}$$
$$14 \div 7 = ?$$

As soon as you're comfortable with addition, subtraction, or whatever, you can switch the equation around a bit:

$$9 + \underline{} = 14$$
$$? \cdot 6 = 18$$

When you stop using underlines and question marks and start using variables such as x to stand for the part of the equation you want you to figure out, bingo! You have an algebra problem:

$$4 + 1 = x$$
$$12 \div x = 3$$
$$x - 13 = 30$$

Four ways to solve algebraic equations

You don't need to call an exterminator just to kill a bug. Similarly, algebra is strong stuff, and you don't always need it to solve an algebraic equation.

Generally speaking, you have four ways to solve algebraic equations such as those I introduce earlier in this chapter:

✔ Eyeballing them (also called inspection, or just looking at the problem to get the answer)

✔ Rearranging them, using inverse operations when necessary

✔ Guessing and checking

✔ Applying algebra

Eyeballing easy equations

You can solve easy problems just by looking at them. For example:

$$5 + x = 6$$

When you look at this problem, you can see that $x = 1$. When a problem is this easy and you can see the answer, you don't need to take any particular trouble to solve it.

Rearranging slightly harder equations

When you can't see an answer just by looking at a problem, sometimes rearranging the problem helps to turn it into one that you can solve using a Big Four operation. This is the method that I use throughout most of this book, especially for solving equations with formulas, like geometry problems and problems involving weights and measures. For example:

$$6x = 96$$

The answer may not jump out at you, but remember that this problem means

$$6 \cdot x = 96$$

You can rearrange this problem using inverse operations, as I show you in Chapter 4:

$$96 \div 6 = x$$

Now, solve the problem by division (long or otherwise) to find that $x = 16$.

Guessing and checking equations

You can solve some equations by guessing an answer and then checking to see whether you're right. For example, suppose you want to solve the following equation:

$$3x + 7 = 19$$

To find out what x equals, start by guessing that $x = 2$. Now, check to see whether you're right by substituting 2 for x in the equation:

$$3(2) + 7 = 6 + 7 = 13 < 19 \quad \text{WRONG!}$$

When $x = 2$, the left side of the equation equals 13, not 19. That was a low guess, so try a higher guess: $x = 5$:

$$3(5) + 7 = 15 + 7 = 22 > 19 \quad \text{WRONG!}$$

This time, your guess was high, so try $x = 4$:

$$3(4) + 7 = 12 + 7 = 19 \quad \text{RIGHT!}$$

With only three guesses, you found that $x = 4$.

Applying algebra to more difficult equations

When an algebraic equation gets hard enough, you find that looking at it and rearranging it just isn't enough to solve it. For example:

$$11x - 13 = 9x + 3$$

You probably can't tell what x equals just by looking at this problem. You also can't solve it just by rearranging it using an inverse operation. And guessing and checking would be very tedious. That's where algebra comes into play.

Algebra is especially useful because you can follow mathematical rules to find your answer. Throughout the rest of this chapter, I show you how to use the rules of algebra to turn tough problems like this one into problems that you can solve.

The Balancing Act: Solving for X

As I show you in the previous section, some problems are too complicated to find out what the variable (usually x) equals just by eyeballing it or rearranging it. For these problems, you need a reliable method for getting the right answer. I call this method the *balancing scale*.

The balancing scale allows you to *solve for x* — that is, find the number that *x* stands for — in a step-by-step process that always works. In this section, I show you how to use the balancing scale method to solve algebraic equations.

Striking a balance

The equal sign in any equation means that both sides balance. To keep that equal sign, you have to maintain that balance. In other words, whatever you do to one side of an equation, you have to do to the other.

For example, here's a balanced equation:

$$\underline{1 + 2 = 3}$$
$$\triangle$$

If you add 1 to one side of the equation, the scale would go out of balance.

$$\underline{1 + 2 + 1 \neq 3}$$
$$\triangle$$

But if you add 1 to *both* sides of the equation, the scale stays balanced:

$$\underline{1 + 2 + 1 = 3 + 1}$$
$$\triangle$$

You can add any number to the equation, as long as you do it to both sides. And in math, *any number* means *x*:

$$1 + 2 + x = 3 + x$$

Remember that *x* is the same wherever it appears in a single equation or problem.

This idea of changing both sides of an equation equally isn't limited to addition. You can just as easily subtract an *x*, or even multiply or divide by *x*, as long as you do the same thing to both sides of the equation:

Subtract: $1 + 2 - x = 3 - x$

Multiply: $(1 + 2)x = 3x$

Divide: $\dfrac{1 + 2}{x} = \dfrac{3}{x}$

Using the balance scale to isolate x

The simple idea of balance is at the heart of algebra, and it enables you to find out what x is in many equations. When you solve an algebraic equation, the goal is to *isolate x* — that is, to get x alone on one side of the equation and some number on the other side. In algebraic equations of middling difficulty, this is a three-step process:

1. Get all constants (non-x terms) on one side of the equation.

2. Get all x-terms on the other side of the equation.

3. Divide to isolate x.

For example, take a look at the following problem:

$$11x - 13 = 9x + 3$$

As you follow the steps, notice how I keep the equation balanced at each step:

1. Get all of the constants on one side of the equation by adding 13 to both sides of the equation:

$$\begin{array}{rcl} 11x - 13 &=& 9x + 3 \\ +\,13 & & +\,13 \\ \hline 11x & =& 9x + 16 \end{array}$$

Because you've obeyed the rules of the balance scale, you know that this new equation is also correct. And now, the only non-x term (16) is on the right side of the equation.

2. Get all of the x-terms on the other side by subtracting $9x$ from both sides of the equation:

$$\begin{array}{rcl} 11x &=& 9x + 16 \\ -\,9x & & -\,9x \\ \hline 2x &=& 16 \end{array}$$

Again, the balance is preserved, so the new equation is correct.

3. Divide by 2 to isolate x:

$$\frac{2x}{2} = \frac{16}{2}$$
$$x = 8$$

To check this answer, you can simply substitute 8 for x in the original equation:

$$11(8) - 13 = 9(8) + 3$$
$$88 - 13 = 72 + 3$$
$$75 = 75 \quad \checkmark$$

This checks out, so 8 is the correct value of x.

Rearranging Equations and Isolating X

When you understand how algebra works like a balance scale, as I show you in the previous section, you can begin to solve more difficult algebraic equations. The basic tactic is always the same: Changing both sides of the equation equally at every step, try to isolate x on one side of the equation.

In this section, I show you how to you put your skills from Chapter 21 to work solving equations. First, I show you how rearranging the terms in an expression is similar to rearranging them in an algebraic equation. Next, I show you how removing parentheses from an equation can help you solve it. Finally, you discover how cross-multiplication is useful for solving algebraic equations with fractions.

Rearranging terms on one side of an equation

Rearranging terms becomes all-important when working with equations. For example, suppose you're working with this equation:

$$5x - 4 = 2x + 2$$

When you think about it, this equation is really two expressions connected with an equal sign. And of course, that's true of *every* equation. That's why everything you find out about expressions in Chapter 21 is useful for solving equations. For example, you can rearrange the terms on one side of an equation. So here's another way to write the same equation:

$$-4 + 5x = 2x + 2$$

And, here's a third way:

$$-4 + 5x = 2 + 2x$$

This flexibility to rearrange terms comes in handy when you're solving equations.

Moving terms to the other side of the equal sign

Earlier in this chapter, I show you how an equation is similar to a balance scale. For example, take a look at Figure 22-1.

Figure 22-1:
Showing
how an
equation is
similar to a
balance
scale.

To keep the scale balanced, if you add or remove anything on one side, you must do the same thing on the other side. For example:

$$2x - 3 = 11$$
$$\underline{-2x \qquad\qquad - 2x}$$
$$-3 = 11 - 2x$$

Now take a look at these two versions of this equation side by side:

$$2x - 3 = 11 \qquad -3 = 11 - 2x$$

In the first version, the term $2x$ is on the left side of the equal sign. In the second, the term $-2x$ is on the right side. This example illustrates an important rule.

When you move any term in an expression to the other side of the equal sign, change its sign (from plus to minus or from minus to plus).

As another example, suppose you're working with this equation:

$$4x - 2 = 3x + 1$$

You have x's on both sides of the equation, so say you want to move the $3x$. When you move the term $3x$ from the right side to the left side, you have to change its sign from plus to minus (technically, you're subtracting $3x$ from both sides of the equation):

$$4x - 2 - 3x = 1$$

After that, you can simplify the expression on the left side of the equation by combining similar terms.

$$x - 2 = 1$$

At this point, you can probably see that $x = 3$ because $3 - 2 = 1$. But, just to be sure, move the -2 term to the right side and change its sign:

$$x = 1 + 2$$

$$x = 3$$

To check this result, substitute a 3 wherever x appears in the original equation:

$$4\,(3) - 2 = 3\,(3) + 1$$

$$12 - 2 = 9 + 1$$

$$10 = 10 \quad \checkmark$$

As you can see, moving terms from one side of an equation to the other can be a big help when you're solving equations.

Removing parentheses from equations

Chapter 21 gives you a treasure trove of tricks for simplifying expressions, and they come in very handy when you're solving equations. One key skill from that chapter is removing parentheses from expressions. This is also indispensable when you're solving equations.

For example, suppose you have the following equation:

$$5x + (6x - 15) = 30 - (x - 7) + 8$$

Your mission is to get all the x terms on one side of the equation and all the constants on the other. As the equation stands, however, x-terms and constants are "locked together" inside parentheses. That is, you can't isolate the x-terms from the constants. So before you can isolate terms, you need to remove the parentheses from the equation.

Recall that an equation is really just two expressions connected by an equal sign. So you can start working with the expression on the left side. In this expression, the parentheses begin with a plus sign (+), so you can just remove them:

$$5x + \underline{6x - 15} = 30 - (x - 7) + 8$$

Now move on to the expression on the right side. This time, the parentheses come right after a minus sign (–). To remove them, change the sign of both terms inside the parentheses: x becomes $-x$, and -7 becomes 7:

$$5x + 6x - 15 = 30 \underline{- x + 7} + 8$$

Bravo! Now you can isolate x terms to your heart's content. Move the $-x$ from the right side to the left, changing it to x:

$$5x + 6x - 15 \underline{+ x} = 30 + 7 + 8$$

Next, move -15 from the left side to the right, changing it to 15:

$$5x + 6x + x = 30 + 7 + 8 \underline{+ 15}$$

Now combine similar terms on both sides of the equation:

$$12x = 30 + 7 + 8 + 15$$
$$12x = 60$$

Finally, get rid of the coefficient 12 by dividing:

$$\frac{12x}{12} = \frac{60}{12}$$
$$x = 5$$

As usual, you can check your answer by substituting 5 into the original equation wherever x appears:

$$5x + (6x - 15) = 30 - (x - 7) + 8$$
$$5(5) + [6(5) - 15] = 30 - (5 - 7) + 8$$
$$25 + (30 - 15) = 30 - (-2) + 8$$
$$25 + 15 = 30 + 2 + 8$$
$$40 = 40 \checkmark$$

Here's one more example:

$$11 + 3(-3x + 1) = 25 - (7x - 3) - 12$$

As in the preceding example, start out by removing both sets of parentheses. This time, however, on the left side of the equation, you have no sign between 3 and $(-3x + 1)$. But again, you can put your skills from Chapter 21 to use. To remove the parentheses, multiply 3 by both terms inside the parentheses:

$$11 - 9x + 3 = 25 - (7x - 3) - 12$$

On the right side, the parentheses begin with a minus sign, so remove the parentheses by changing all the signs inside the parentheses:

$$11 - 9x + 3 = 25 - 7x + 3 - 12$$

Now, you're ready to isolate the x terms. I do this in one step, but take as many steps as you like:

$$-9x + 7x = 25 + 3 - 12 - 11 - 3$$

At this point, you can combine similar terms:

$$-2x = 2$$

To finish up, divide both sides by -2:

$$x = -1$$

As always, check your solution using substitution:

$$11 + 3(-3x + 1) = 25 - (7x - 3) - 12$$
$$11 + 3[\underline{-3(-1)} + 1] = 25 - [\underline{7(-1)} - 3] - 12$$

All the variables are gone now, so remember the order of precedence rules from Chapter 5. Begin with the multiplication inside the parentheses, which I've underlined:

$$11 + 3(3 + 1) = 25 - (-7 - 3) - 12$$

Now you can simplify what's inside each set of parentheses:

$$11 + 3(4) = 25 - (-10) - 12$$

And now you can remove the parentheses and complete the check:

$$11 + 12 = 25 + 10 - 12$$
$$23 = 23 \checkmark$$

Copy this example and work through it a few times with the book closed.

Cross-multiplication

In algebra, cross-multiplication helps to simplify equations by removing unwanted fractions (and, honestly, when are fractions ever wanted?). As I discuss in Chapter 9, you can use cross-multiplication to find out whether two fractions are equal. For example, here are two equal fractions:

$$\frac{2}{4} = \frac{3}{6}$$

When you cross-multiply them, you multiply the numerator of one fraction by the denominator of the other:

$$2(6) = 3(4)$$
$$12 = 12 \checkmark$$

But suppose you want to solve this algebra equation:

$$\frac{x}{2x - 2} = \frac{2x + 3}{4x}$$

This equation looks hairy. You can't do the division or cancel anything out because the fraction on the left has two terms in the denominator, and the fraction on the right has two terms in the numerator (see Chapter 21 for info on dividing algebraic terms). However, an important piece of information that you have is that the fraction $\frac{x}{2x - 2}$ equals the fraction $\frac{2x + 3}{4x}$. So if you cross-multiply these two fractions, you get two results that are also equal:

$$x(4x) = (2x + 3)(2x - 2)$$

At this point, you have something you know how to work with. The left side is easy:

$$4x^2 = (2x + 3)(2x - 2)$$

The right side requires a bit of FOILing (flip to Chapter 21 for details):

$$4x^2 = 4x^2 - 4x + 6x - 6$$

Now all the parentheses are gone, so you can isolate the x terms. Because most of these terms are already on the right side of the equation, isolate them on that side:

$$6 = 4x^2 - 4x + 6x - 4x^2$$

Combining like terms gives you a very pleasant surprise:

$$6 = 2x$$

The two x^2 terms cancel each other out. You may be able to eyeball the correct answer, but here's how to finish:

$$\frac{6}{2} = \frac{2x}{2}$$
$$x = 3$$

To check your answer, substitute 3 back into the original equation:

$$\frac{3}{2(3) - 2} = \frac{2(3) + 3}{4(3)}$$
$$\frac{3}{6 - 2} = \frac{6 + 3}{12}$$
$$\frac{3}{4} = \frac{9}{12}$$
$$\frac{3}{4} = \frac{3}{4}$$

This checks out, so the answer $x = 3$ is correct.

Chapter 23

Putting Mr. X to Work: Algebra Word Problems

In This Chapter
▶ Solving algebra word problems in simple steps
▶ Working through solving an algebra word problem
▶ Choosing variables
▶ Using charts

*W*ord problems that require algebra are among the toughest problems that students face — and the most common. Teachers just love algebra word problems because they bring together a lot of what you know, such as solving algebra equations (Chapters 21 and 22) and turning words into numbers (see Chapters 6, 13, and 18). And standardized tests virtually always include these types of problems.

In this chapter, I show you a five-step method for using algebra to solve word problems. Then, I give you a bunch of examples that take you through all five steps.

Along the way, I give you some important tips that can make solving word problems easier. First, I show you how to choose a variable that makes your equation as simple as possible. Next, I give you practice organizing information from the problem into a chart. By the end of this chapter, you should have a solid understanding of how to solve a wide variety of algebra word problems.

Solving Algebra Word Problems in Five Steps

Everything from Chapters 21 and 22 comes into play when you use algebra to solve word problems, so if you feel a little shaky on solving algebraic equations, flip back to those chapters for some review.

Throughout this section, I use the following word problem as an example:

> In three days, Alexandra sold a total of 31 tickets to her school play. On Tuesday, she sold twice as many tickets as on Wednesday. And on Thursday, she sold exactly 7 tickets. How many tickets did Alexandra sell on each day, Tuesday through Thursday?

Organizing the information in an algebra word problem using a chart or picture is usually helpful. Here's what I came up with:

Tuesday:	twice as many as on Wednesday
Wednesday:	?
Thursday:	7
Total:	31

At this point, all the information is in the chart, but the answer still may not be jumping out at you. In this section, I outline a step-by-step method that enables you to solve this problem and much harder ones as well.

Here are the five steps for solving most algebra word problems:

1. **Declare a variable.**
2. **Set up the equation.**
3. **Solve the equation.**
4. **Answer the question that the problem asks.**
5. **Check your answer.**

Declaring a variable

As you know from Chapter 21, a variable is a letter that stands for a number. Most of the time, you don't find the variable x (or any other variable, for that matter) in a word problem. That doesn't mean you don't need algebra to solve the problem. It just means that you're going to have to put *x* into the problem yourself and decide what it stands for.

When you *declare a variable,* you say what that variable means in the problem you're solving.

Here are some examples of variable declarations:

Let *m* = the number of dead mice that the cat dragged into the house.

Let *p* = the number of times Marianne's husband promised to take out the garbage.

Let *c* = the number of complaints Arnold received after he painted his garage door purple.

In each case, you take a variable (*m*, *p*, or *c*) and give it a meaning by attaching it to a number.

Notice that the earlier chart for the sample problem has a big question mark next to *Wednesday*. This question mark stands for *some number,* so you may want to declare a variable that stands for this number. Here's how you do it:

Let *w* = the number of tickets that Alexandra sold on Wednesday.

Whenever possible, choose a variable with the same initial as what the variable stands for. This practice makes remembering what the variable means a lot easier, which will help you later in the problem.

For the rest of the problem, every time you see the variable *w*, keep in mind that it stands for the number of tickets that Alexandra sold on Wednesday.

Setting up the equation

After you have a variable to work with, you can go through the problem again and find other ways to use this variable. For example, Alexandra sold twice as many tickets on Tuesday as on Wednesday, so she sold $2w$ tickets on Tuesday. Now you have a lot more information to fill in the chart:

Tuesday:	Twice as many as on Wednesday	$2w$
Wednesday:	?	w
Thursday:	7	7
Total:	31	31

You know that the total number of tickets, or the sum of the tickets she sold on Tuesday, Wednesday, and Thursday, is 31. With the chart filled in like that, you're ready to set up an equation to solve the problem:

$$2w + w + 7 = 31$$

Solving the equation

After you set up an equation, you can use the tricks from Chapter 22 to solve the equation for w. Here's the equation one more time:

$$2w + w + 7 = 31$$

For starters, remember that $2w$ really means $w + w$. So on the left, you know you really have $w + w + w$, or $3w$; you can simplify the equation a little bit as follows:

$$3w + 7 = 31$$

The goal at this point is to try to get all the terms with w on one side of the equation and all the terms without w on the other side. So on the left side of the equation, you want to get rid of the 7. The inverse of addition is subtraction, so subtract 7 from both sides:

$$
\begin{array}{r}
3w + 7 = 31 \\
\underline{-7 \quad -7} \\
3w \quad\; = 24
\end{array}
$$

You now want to isolate w on the left side of the equation. To do this, you have to undo the multiplication by 3, so divide both sides by 3:

$$\frac{3w}{3} = \frac{24}{3}$$
$$w = 8$$

Answering the question

You may be tempted to think that after you've solved the equation, you're done. But you still have a bit more work to do. Look back at the problem, and you see that it asks you this question:

How many tickets did Alexandra sell on each day, Tuesday through Thursday?

At this point, you have some information that can help you solve the problem. The problem tells you that Alexandra sold 7 tickets on Thursday. And because $w = 8$, you now know that she sold 8 tickets on Wednesday. And on Tuesday, she sold twice as many on Wednesday, so she sold 16. So Alexandra sold 16 tickets on Tuesday, 8 on Wednesday, and 7 on Thursday.

Checking your work

To check your work, compare your answer to the problem, line by line, to make sure every statement in the problem is true:

> In three days, Alexandra sold a total of 31 tickets to her school play.

That's correct, because $16 + 8 + 7 = 31$.

> On Tuesday, she sold twice as many tickets as on Wednesday.

Correct, because she sold 16 tickets on Tuesday and 8 on Wednesday.

> And on Thursday, she sold exactly 7 tickets.

Yep, that's right, too, so you're good to go.

Choosing Your Variable Wisely

Declaring a variable is simple, as I show you earlier in this chapter, but you can make the rest of your work a lot easier when you know how to choose your variable wisely. Whenever possible, choose a variable so that the equation you have to solve has no fractions, which are much more difficult to work with than whole numbers.

For example, suppose you're trying to solve this problem:

> Irina has three times as many clients as Toby. If they have 52 clients all together, how many clients does each person have?

The key sentence in the problem is "Irina has *three times as many* clients as Toby." It's significant because it indicates a relationship between Irina and Toby that's based on either *multiplication or division*. And to avoid fractions, you want to avoid division wherever possible.

Whenever you see a sentence that indicates you should use either multiplication or division, choose your variable to represent the *smaller* number. In this case, Toby has fewer clients than Irina, so choosing t as your variable is the smart move.

Suppose you begin by declaring your variable as follows:

Let t = the number of clients that Toby has.

Then, using that variable, you can make this chart:

Irina	$3t$
Toby	t

No fraction! Now, to solve this problem, set up this equation:

Irina + Toby = 52

Plug in the values from the chart:

$3t + t = 52$

Now you can solve the problem easily using what I show you in Chapter 22:

$4t = 52$

$t = 13$

Toby has 13 clients, so Irina has 39. To check this result — which I recommend highly earlier in this chapter! — note that $13 + 39 = 52$.

Now suppose that instead, you take the opposite route and decide to declare a variable as follows:

Let i = the number of clients that Irina has.

Given that variable, you'd have to represent Toby's clients using the fraction $\frac{i}{3}$, which leads to the same answer but a *lot* more work.

Solving More-Complex Algebraic Problems

Algebra word problems become more complex when the number of people or things that you need to find out increases. In this section, the complexity increases from two or three people to four and then five. When you're done, you should feel comfortable solving algebra word problems of significant difficulty.

Charting four people

As in the previous section, a chart can help you organize information so you don't get confused. Here's a problem that involves four people:

> Alison, Jeremy, Liz, and Raymond participated in a canned goods drive at work. Liz donated three times as many cans as Jeremy, Alison donated twice as many as Jeremy, and Raymond donated 7 more than Liz. Together, the two women donated two more cans than the two men. How many cans did the four people donate all together?

The first step, as always, is declaring a variable. Remember that to avoid fractions, you want to declare a variable based on the person who brought in the fewest cans. But Liz donated more cans than Jeremy, and so did Alison. Furthermore, Raymond donated more cans than Liz. So, because Jeremy donated the fewest cans, declare your variable as follows:

Let j = the number of cans that Jeremy donated

Now you can set up your chart as follows:

Jeremy	j
Liz	$3j$
Alison	$2j$
Raymond	Liz + 7 = $3j + 7$

This looks good because, as expected, there are no fractional amounts in the chart. The next sentence tells you that the women donated two more cans than the men, so make a word problem, as I show you in Chapter 6:

Liz + Alison = Jeremy + Raymond + 2

You can now substitute into this equation as follows:

$3j + 2j = j + 3j + 7 + 2$

With your equation set up, you're ready to solve. First, isolate the algebraic terms:

$3j + 2j - j - 3j = 7 + 2$

Combine similar terms:

$j = 9$

Almost without effort, you've solved the equation, so you know that Jeremy donated 9 cans. With this information, you can go back to the chart, plug in 9 for j, and find out how many cans the other children donated: Liz donated 27, Alison donated 18, and Raymond donated 34. Finally, you can add up these numbers to conclude that the four people donated 88 cans altogether.

To check the numbers, read through the problem and make sure they work at every point in the story. For example, together Liz and Alison donated 45 cans and Jeremy and Raymond donated 43, so the women really did donate 2 more cans than the men.

Crossing the finish line with five people

Here's one final example, the most difficult in this chapter, in which you have five people to work with.

> Five friends are keeping track of how many miles they run. So far this month, Mina has run 12 miles, Suzanne has run 3 more miles than Jake, and Kyle has run twice as far as Victor. But tomorrow, after they all complete a 5-mile run, Jake will have run as far as Mina and Victor combined, and the whole group will have run 174 miles. How far has each person run so far?

The most important thing to notice in this problem is that there are two sets of numbers: the miles that all five people have run up to *today* and their mileage including *tomorrow*. And each person's mileage tomorrow will be five miles greater than his or her mileage today. Here's how to set up a chart:

	Today	*Tomorrow (Today + 5)*
Jake		
Kyle		
Mina		
Suzanne		
Victor		

With this chart, you're off to a good start to solve this problem. Next, look for that statement early in the problem that connects two people by either multiplication or division. Here it is:

> . . . Kyle has run *twice as far* as Victor.

Because Victor has run fewer miles than Kyle, declare your variable as follows:

> Let v = the number of miles that Victor has run up to *today*.

Notice that I added the word *today* to the declaration to be very clear that I'm talking about Victor's miles *before* the 5-mile run tomorrow.

At this point, you can begin filling in the chart:

	Today	*Tomorrow (Today + 5)*
Jake		
Kyle	$2v$	$2v + 5$
Mina	12	17
Suzanne		
Victor	v	$v + 5$

As you can see, I left out the information about Jake and Suzanne because I can't represent it using the variable v. I've also begun to fill in the *Tomorrow* column by adding 5 to my numbers in the *Today* column.

Now I can move on to the next statement in the problem:

> But tomorrow . . . Jake will have run as far as Mina and Victor combined. . . .

I can use this to fill in Jake's information:

	Today	*Tomorrow (Today + 5)*
Jake	$17 + v$	$17 + v + 5$
Kyle	$2v$	$2v + 5$
Mina	12	17
Suzanne		
Victor	v	$v + 5$

In this case, I first filled in Jake's *tomorrow* distance ($17 + v + 5$) and then subtracted 5 to find out his *today* distance. Now I can use the information that, today, Suzanne has run 3 more miles than Jake:

	Today	*Tomorrow (Today + 5)*
Jake	$17 + v$	$17 + v + 5$
Kyle	$2v$	$2v + 5$
Mina	12	17
Suzanne	$17 + v + 3$	$17 + v + 8$
Victor	v	$v + 5$

With the chart filled in like this, you can begin to set up your equation. First, set up a word equation as follows:

Jake tomorrow + Kyle tomorrow + Mina tomorrow + Suzanne tomorrow + Victor tomorrow = 174

Now, just substitute information from the chart into this word equation to set up your equation:

$$17 + v + 5 + 2v + 5 + 17 + 17 + v + 8 + v + 5 = 174$$

As always, begin solving by isolating the algebraic terms:

$$v + 2v + v + v = 174 - 17 - 5 - 5 - 17 - 17 - 8 - 5$$

Next, combine similar terms:

$$5v = 100$$

Finally, to get rid of the coefficient in the term $5v$, divide both sides by 5:

$$\frac{5v}{5} = \frac{100}{5}$$
$$v = 20$$

You now know that Victor's total distance up to *today* is 20 miles. With this information, you substitute 20 for v and fill in the chart as follows:

	Today	*Tomorrow (Today + 5)*
Jake	37	42
Kyle	40	45
Mina	12	17
Suzanne	40	45
Victor	20	25

The *Today* column contains the answers to the question that the problem asks. To check this solution, make sure that every statement in the problem is true. For example, tomorrow the five people will have run a total of 174 miles because

$$42 + 45 + 17 + 45 + 25 = 174$$

Copy down this problem, close the book, and work through it for practice.

Part VI
The Part of Tens

The 5th Wave   By Rich Tennant

Taking Algebra Tests Was Particularly Embarrassing For Mr. Ed Because of His Inability to Keep His Calculations to Himself.

In this part . . .

*J*ust for fun, this part of the book includes a couple of top-ten lists on math-related topics. I show you some important math concepts you should remember not to forget. And I list some very important sets of numbers.

Chapter 24

Ten Key Math Concepts You Shouldn't Ignore

....

In This Chapter

▶ Appreciating how simple concepts are indispensable in math

▶ Seeing why π and prime numbers are important

▶ Understanding the importance of sets and functions

▶ Looking at more-advanced concepts

....

Math itself is one big *concept*, and it's chock full of so many smaller concepts that no one person, however long he or she studies, can possibly understand them all.

But inside all this, certain concepts get so much airplay that in my humble opinion, they make the Math Hall of Fame. Each of these ideas changed not only math but also the way people thought of the world. Knowing them can change your world, too — or at least give you a broader perspective of what math is all about.

So here's my list of the ten most important concepts in mathematics.

Getting Set with Sets

A *set* is a collection of objects. The objects, called *elements* of the set, can be tangible (shoes, bobcats, people, jellybeans, and so forth) or intangible (fictional characters, ideas, numbers, and the like).

Sets are such a simple and flexible way of organizing the world that you can define all of math in terms of them. How mathematicians do this is very complex, but a basic understanding of sets isn't difficult and is part of most math education. See Chapter 20 for more on set theory and Chapter 25 for some significant number sets.

Playing with Prime Numbers

A *prime number* is any counting number that has exactly two divisors (numbers that divide into it evenly) — 1 and the number itself. Here are the first ten prime numbers:

2   3   5   7   11   13   17   19   23   29 ...

Prime numbers go on forever — that is, the list is infinite.

Beyond this, prime numbers are, in an important sense, the elements from which all other numbers can be built. Every counting number greater than 1, no matter how large, can be written as the unique product of prime numbers.

Far more than just a curious bit of trivia, the uniqueness of each number's prime factors goes by a much more important name: the Fundamental Theorem of Arithmetic. See Chapters 1 and 7 for more on prime numbers.

Zero: Much Ado about Nothing

Zero may look like a big nothing, but it's actually one of the greatest inventions of all time. Like all inventions, it didn't exist until someone thought of it. The Greeks and Romans, who knew so much about math and logic, knew nothing about zero. The number systems they used had no way to express, for example, how many olive trees you had left when you started with three and an angry neighbor cut down all three of them.

The concept of zero as a number arose independently in several different places. In South America, the number system that the Mayans used included a symbol for zero. And the Hindu-Arabic system used throughout most of the world today developed from an earlier Arabic system that used zero as a placeholder. (See Chapter 2 for more on the use of zero as a placeholder.)

In fact, *zero* isn't really nothing — it's simply a way to express *nothing* mathematically. And that's really something.

Going Greek: Pi (π)

The symbol π (pi — pronounced *pie*) is a Greek letter that stands for the ratio of the circumference of a circle to its diameter (see Chapter 16 for the scoop on circles). Here's the approximate value of π:

$$\pi \approx 3.1415926535 \ldots$$

Although π is just a number — or, in algebraic terms, a constant — it's important for several reasons:

✔ Geometry just wouldn't be the same without it. Circles are one of the most basic shapes in geometry, and you need π to measure the area and the circumference of a circle. So if aliens land in your cornfield and you want to measure the resulting crop circles, or if you just want to know the area of your round kitchen table, π can come in handy.

✔ Pi is an *irrational number,* which means that no fraction that equals it exactly exists. Beyond this, π is a *transcendental number,* which means that it's never the value of x in a polynomial equation (the most basic type of algebraic equation). Thus, even though π emerges from a very simple operation (measuring a circle), it contains a deep complexity that numbers such as 0, 1, –1, ½, and even $\sqrt{2}$ don't share. (See Chapter 25 for details on irrational and transcendental numbers.)

✔ Pi is everywhere in math. It shows up constantly (no pun intended) where you'd least expect it. One example is trigonometry, the study of triangles. Triangles obviously aren't circles, but trig uses circles to measure the size of angles, and you can't swing a compass without hitting π.

On the Level: Equal Signs and Equations

Almost everyone takes the humble equal sign (=) for granted. It's so common in math that it goes virtually unnoticed. But the fact that the equal sign shows up practically everywhere only adds weight to the idea that the concept of *equality* — an understanding of when one thing is mathematically the same as another — is one of the most important math concepts ever created.

A mathematical statement with an equal sign is an *equation.* The equal sign links two mathematical expressions that have the same value. The power of math lies in this linkage. That's why nearly everything in math involves equations. On their own, expressions are limited in their usefulness. The equal sign provides a powerful way to connect expressions, which allows scientists to connect ideas in new ways.

For example, for thousands of years, energy and matter were seen as separate and unrelated. Einstein's famous equation — $E = mc^2$ — links an expression that represents energy with one that represents matter. The result is a radically altered view of the universe.

For details on how the concepts of equality and balance play out in algebra, flip to Chapter 22.

On the Grid: The Cartesian Graph

The Cartesian graph (also called the Cartesian coordinate system) is the fancy name for the good old-fashioned graph that I discuss in Chapter 17. It was invented by French philosopher and mathematician René Descartes.

What's so special about the Cartesian graph? Before the graph was invented, algebra and geometry were studied for centuries as two separate and unrelated areas of math. Algebra was exclusively the study of equations (see Part V), and geometry was solely the study of figures on the plane or in space (see Chapter 16).

Descartes' invention of the graph brought algebra and geometry together. The result was *analytic geometry,* a new mathematics that not only merged the ancient sciences of algebra and geometry but also brought greater clarity to both. Now, you can draw solutions to equations that include the variables x and y as points, lines, circles, and other geometric shapes on a graph.

In and Out: Relying on Functions

A *function* is a mathematical machine that takes in one number (called the *input*) and gives back exactly one other number (called the *output*). It's kind of like a blender because what you get out of it depends on what you put into it: When you put in ice cream, you get a milkshake. When you put in fruit, you get a fruit smoothie. When you put in your cell phone, you get a heinously damaged cell phone.

Suppose I invent a function called PlusOne that adds 1 to any number. So when you input the number 2, the number that gets outputted is 3:

PlusOne (2) = 3

Similarly, when you input the number 100, the number that gets outputted is 101:

PlusOne(100) = 101

As you can see, whenever you input an even number, the function PlusOne outputs an odd number. And this will happen for every even number. Thus, this function maps the set of even numbers to the set of odd numbers.

This process may look rather simplistic, but as with sets, the simplicity of functions gives them their power. Functions allow mathematicians — and a host of other people, such as computer programmers, statisticians, biologists, economists, and psychologists — to perceive a complex world in a mathematical way.

Functions get a lot of play as you move forward in algebra. For now, just remember that a function takes an input and gives you an output. For a deeper look at functions, see *Algebra For Dummies* by Mary Jane Sterling (Wiley).

Exploring the Infinite

The very word *infinity* commands great power. So does the symbol for infinity (∞). How big is infinity? Here's a common answer: If you were to count all the grains of sand on all the beaches in the world and then do the same thing on every planet in our galaxy, by the time you were done counting, you'd be no closer to infinity than you are right now. Yessiree, that's big. (Don't you wish you'd gone sightseeing on all those other planets instead?)

In fact, infinity isn't a number at all. *Infinity,* beyond any classification of size or number, is the very quality of endlessness. And yet, mathematicians have tamed infinity to a great extent.

In his invention of calculus, Sir Isaac Newton introduced the concept of a *limit,* which allows you to calculate what happens to numbers as they get very large and approach infinity. And in his transfinite mathematics, Georg Cantor proved that the very infinity I just described is only the very smallest of a set of larger infinities. And here's the kicker: This set of larger and larger infinities itself is *infinite.* (See Chapter 25 for more on Cantor's transfinite numbers.)

The Real Number Line

The number line has been around for a very long time, and it's one of the first visual aids that teachers use to teach kids about numbers. Every point on the number line stands for a number. Well, okay, that sounds pretty obvious, but strange to say, this concept wasn't *fully* understood for thousands of years.

The Greek philosopher Zeno of Elea posed this problem, called *Zeno's Paradox:* In order to walk across the room, you have to first walk half the distance (½) across the room. Then you have to go half the remaining distance (¼). After that, you have to go half the distance that still remains (⅛). This pattern continues forever:

½   ¼   ⅛   ¹⁄₁₆   ¹⁄₃₂   ¹⁄₆₄   ¹⁄₁₂₈   ¹⁄₂₅₆ ...

So you can *never* get to the other side of the room.

Obviously, in the real world, you can and do walk across rooms all the time. But from the standpoint of math, Zeno's Paradox and other similar paradoxes remained unanswered for about 2,000 years.

The basic problem was this: All the fractions listed in the preceding sequence are between 0 and 1 on the number line. And there are an infinite number of them. But how can you have an *infinite* number of numbers in a *finite* space?

Mathematicians of the 19th century — Augustin Cauchy, Richard Dedekind, Karl Weierstrass, and Georg Cantor foremost among them — solved this paradox. The result was *real analysis,* the advanced mathematics of the real number line.

The Imaginary Number i

The *imaginary numbers* are a set of numbers not found on the real number line. If that idea sounds unbelievable — where else would they be? — don't worry: For thousands of years, mathematicians didn't believe in them, either. But real-world applications in electronics, particle physics, and many other areas of science have turned skeptics into believers. So if your summer plans include wiring your secret underground lab or building a flux capacitor for your time machine — or maybe just studying to get a degree in electrical engineering — you'll find that imaginary numbers are too useful to be ignored.

See Chapter 25 for info on imaginary and complex numbers.

Chapter 25

Ten Important Number Sets You Should Know

. . . .

In This Chapter

▶ Identifying counting numbers, integers, rational numbers, and real numbers

▶ Discovering imaginary and complex numbers

▶ Looking at how transfinite numbers represent higher levels of infinity

. . . .

*T*he more you find out about numbers, the stranger they become. When you're working with just the counting numbers and a few simple operations, numbers seem to develop a landscape all their own. The terrain of this landscape starts out uneventful, but as you introduce other sets, it soon turns surprising, shocking, and even mind-blowing. In this chapter, I take you on a mind-expanding tour of ten sets of numbers.

I start with the very familiar and comfy counting numbers. I continue with the integers (positive and negative counting numbers and 0), the rational numbers (integers and fractions), and real numbers (all numbers on the number line). I also take you on a few side routes along the way. The tour ends with the bizarre and almost unbelievable transfinite numbers. And in a way, the transfinite numbers bring you back to where you started at the very beginning — with the counting numbers.

Each of these sets of numbers serves a different purpose, some familiar (such as accounting or carpentry), some scientific (such as electronics and physics), and a few purely mathematical. Enjoy the ride!

Counting on Counting (Or Natural) Numbers

The *counting numbers* — also called the *natural numbers* — are probably the first numbers you ever encountered. They start with 1 and go up from there:

1, 2, 3, 4, 5, 6, 7, 8, 9, 10, 11, 12, ...

The three dots (or ellipsis) at the end tell you that the sequence of numbers goes on forever — in other words, it's infinite.

The counting numbers are useful for keeping track of tangible objects: stones, chickens, cars, cellphones — anything that you can touch and that you don't plan to cut into pieces.

The set of counting numbers is *closed* under both addition and multiplication. That is, if you add or multiply any two counting numbers, the result is also a counting number. But the set isn't closed under subtraction or division. For example, if you subtract 2 – 3, you get –1, which is a negative number rather than a counting number. And if you divide 2 ÷ 3, you get ⅔, which is a fraction.

If you place 0 in the set of counting numbers, you get the set of *whole numbers*.

Identifying Integers

The set of *integers* includes the counting numbers (see the preceding section), the negative counting numbers, and 0:

..., –6, –5, –4, –3, –2, –1, 0, 1, 2, 3, 4, 5, 6, ...

The dots, or ellipses, at the beginning and the end of the set tell you that the integers are infinite in both the positive and negative directions.

Because the integers include the negative numbers, you can use them to keep track of anything that can potentially involve debt. In today's culture, that's usually money. For example, if you have $100 in your checking account and write a check for $120, you find that your new balance will drop to –$20 (not counting any fees that the bank charges!).

The set of integers is *closed* under addition, subtraction, and multiplication. In other words, if you add, subtract, or multiply any two integers, the result is also an integer. But the set isn't closed under division. For example, if you divide the integer –2 by the integer 5, you get the fraction –⅖, which isn't an integer.

Knowing the Rationale behind Rational Numbers

The *rational numbers* include the integers (see the preceding section) and all the fractions between the integers. Here, I list only the rational numbers from –1 to 1 whose denominators (bottom numbers) are positive numbers less than 5:

... –1 ... –¾ ... –⅔ ... –½ ... –⅓ ... –¼ ... 0 ... ¼ ... ⅓ ... ½ ... ⅔ ... ¾ ... 1 ...

The ellipses tell you that between any pair of rational numbers are an infinite number of other rational numbers — a quality called the *infinite density* of rational numbers.

Rational numbers are commonly used for measurement in which precision is important. For example, a ruler wouldn't be much good if it were to measure length only to the nearest inch. Most rulers measure length to the nearest 1/16 of an inch, which is close enough for most purposes. Similarly, measuring cups, scales, precision clocks, and thermometers that allow you to make measurements to a fraction of a unit also use rational numbers. (See Chapter 15 for more on units of measurement.)

The set of rational numbers is closed under the Big Four operations. That is, if you take any two rational numbers and add, subtract, multiply, or divide them, the result is always another rational number.

Making Sense of Irrational Numbers

In a sense, the irrational numbers are a sort of catchall; every number on the number line that isn't rational is irrational.

By definition, no *irrational number* can be represented as a fraction; nor can an irrational number be represented as either a terminating decimal or a repeating decimal (see Chapter 11 for more about these types of decimals).

Instead, an irrational number can be approximated only as a *non-terminating, non-repeating decimal:* the string of numbers after the decimal point goes on forever without creating a pattern.

The most famous example of an irrational number is π, which represents the circumference of a circle with a diameter of 1 unit. Another common irrational number is $\sqrt{2}$, which represents the diagonal distance across a square with a side of 1 unit. In fact, all square roots of non-square numbers (such as $\sqrt{3}$, $\sqrt{5}$, and so forth) are irrational numbers.

Irrational numbers fill out the spaces in the real number line. (The *real number line* is just the number line you're used to, but it's continuous; it has no gaps so that every point is paired with a number.) These numbers are used in many cases in which you need not just a high level of precision, as with the rational numbers, but the *exact* value of a number that can't be represented as a fraction.

Irrational numbers come in two varieties: *algebraic numbers* and *transcendental numbers.* I discuss both of these types of numbers in the sections that follow.

Absorbing Algebraic Numbers

To understand *algebraic numbers,* you need a little information about polynomial equations. A *polynomial equation* is an algebraic equation that meets the following conditions:

- ✔ Its operations are limited to addition, subtraction, and multiplication. In other words, you don't have to divide by a variable.

- ✔ Its variables are raised only to positive, whole-number exponents.

You can find out more about polynomials in *Algebra For Dummies* by Mary Jane Sterling (Wiley). Here are some polynomial equations:

$$2x + 14 = (x + 3)^2$$
$$2x^2 - 9x - 5 = 0$$

Every algebraic number shows up as the solution of at least one polynomial equation. For example, suppose you have the following equation:

$$x^2 = 2$$

You can solve this equation as $x = \sqrt{2}$. Thus, $\sqrt{2}$ is an algebraic number whose approximate value is 1.4142135623... (see Chapter 4 for more information on square roots).

Moving through Transcendental Numbers

A *transcendental number,* in contrast to an algebraic number (see the preceding section), is *never* the solution of a polynomial equation. Like the irrational numbers, transcendental numbers are also a sort of catchall: Every number on the number line that isn't algebraic is transcendental.

The best known transcendental number is π, whose approximate value is 3.1415926535... . Its uses begin in geometry but extend to virtually all areas of mathematics. (See Chapters 16 and 24 for more on π.)

Other important transcendental numbers come about when you study *trigonometry,* the math of right triangles. Sines, cosines, tangents, and other trigonometric functions are often transcendental numbers.

Another important transcendental number is *e,* whose approximate value is 2.7182818285... . The number *e* is the base of the natural logarithm, which you probably won't use until you get to pre-calculus or calculus. People use *e* to do problems on compound interest, population growth, radioactive decay, and the like.

Getting Grounded in Real Numbers

The set of *real numbers* is the set of all rational and irrational numbers (see the earlier sections). The real numbers comprise every point on the number line.

Like the rational numbers (see "Knowing the Rationale behind Rational Numbers," earlier in this chapter), the set of real numbers is closed under the Big Four operations. That is, if you take any two real numbers and add, subtract, multiply, or divide them, the result is always another real number.

Trying to Imagine Imaginary Numbers

An *imaginary number* is any real number multiplied by $\sqrt{-1}$.

To understand what's so strange about imaginary numbers, it helps to know a bit about square roots. The *square root* of any number is a second number that, when multiplied by itself, gives you the first number. For example, the square root of 9 is 3 because $3 \cdot 3 = 9$. And the square root of 9 is also -3 because $-3 \cdot -3 = 9$. (See Chapter 4 for more on square roots and multiplying negative numbers.)

The problem with finding $\sqrt{-1}$ is that it isn't on the real number line (because it isn't in the set of real numbers). If it were on the real number line, it would be a positive number, a negative number, or 0. But when you multiply any positive number by itself, you get a positive number. And when you multiply any negative number by itself, you also get a positive number. Finally, when you multiply 0 by itself, you get 0.

If $\sqrt{-1}$ isn't on the real number line, where is it? That's a good question. For thousands of years, mathematicians believed that the square root of a negative number was simply meaningless. They banished it to the mathematical non-place called *undefined,* which is the same place where they kept fractions with a denominator of 0. In the 19th century, however, mathematicians began to find these numbers useful and found a way to incorporate them into the rest of math.

Mathematicians designated $\sqrt{-1}$ with the symbol *i*. Because it didn't fit onto the real number line, *i* got its own number line, which looks a lot like the real number line. Figure 25-1 shows some numbers that form the imaginary number line.

Figure 25-1:
Numbers
on the
imaginary
number line.

Even though these numbers are called imaginary, mathematicians today consider them no less real than the real numbers. And the scientific application of imaginary numbers to electronics and physics has verified that these numbers are more than just figments of someone's imagination.

Grasping the Complexity of Complex Numbers

A *complex number* is any real number (see "Getting Grounded in Real Numbers," earlier in this chapter) plus or minus an imaginary number (see the preceding section). Here are some examples:

$$1 + i \qquad 5 - 2i \qquad -100 + 10i$$

Getting inside subsets

Many sets of numbers actually fit inside other sets. Mathematicians call these nesting sets *subsets*. For instance, the set of integers is called *Z* for short. Because the set of counting or natural numbers (represented as *N*) is completely contained within the set of integers, *N* is a subset, or part, of *Z*.

The set of rational numbers is called *Q*. Because the set of integers is completely contained within the set of rational numbers, *N* and *Z* are both subsets of *Q*.

R stands for the set of real numbers. Because the set of rational numbers is completely contained within the set of real numbers, *N, Z,* and *Q* are all subsets of *R*.

The set of complex numbers is called *C*. Because the set of real numbers is completely contained within the set of complex numbers, *N, Z, Q,* and *R* are all subsets of *C*.

The symbol $\subset$ means *is a subset of* (see Chapter 20 for details on set notation). So here's how the sets fit inside each other:

$$N \subset Z \subset Q \subset R \subset C$$

You can turn any real number into a complex number by just adding $0i$ (which equals 0):

$$3 = 3 + 0i \qquad -12 = -12 + 0i \qquad 3.14 = 3.14 + 0i$$

These examples show you that the real numbers are just a part of the larger set of complex numbers.

Like the rational numbers and real numbers (check out the sections earlier in this chapter), the set of complex numbers is closed under the Big Four operations. That is, if you take any two complex numbers and add, subtract, multiply, or divide them, the result is always another complex number.

Going beyond the Infinite with Transfinite Numbers

The *transfinite numbers* are a set of numbers representing different levels of infinity. Consider this for a moment: The counting numbers (1, 2, 3, ...) go on forever, so they're infinite. But there are *more* real numbers than counting numbers.

In fact, the real numbers are *infinitely more infinite* than the counting numbers. Mathematician Georg Cantor proved this fact. He also proved that for every level of infinity, you can find another level that's even higher. He called these numbers *transfinite* because they transcend, or go beyond, what you think of as infinite.

The lowest transfinite number is $\aleph_0$ (aleph null), which equals the number of elements in the set of counting numbers. Because the counting numbers are infinite, the familiar symbol for infinity (∞) and $\aleph_0$ mean the same thing.

The next transfinite number is $\aleph_1$ (aleph one), which equals the number of elements in the set of real numbers. This is a higher order of infinity than ∞.

The sets of integers, rational, and algebraic numbers all have $\aleph_0$ elements. And the sets of irrational, transcendental, imaginary, and complex numbers all have $\aleph_1$ elements.

Higher levels of infinity exist, too. Here's the set of transfinite numbers:

$$\aleph_0, \ \aleph_1, \ \aleph_2, \ \aleph_3, \ \ldots$$

The ellipsis tells you that the sequence of transfinite numbers goes on forever — in other words, that it's infinite. As you can see, on the surface, the transfinite numbers look similar to the counting numbers (in the first section of this chapter). That is, the set of transfinite numbers has $\aleph_0$ elements.

Index

• G •

• H •

● *T* ●

• Y •

• Z •